PIC in Practice

D0378126

PIC in Practice

A Project-Based Approach

D. W. Smith

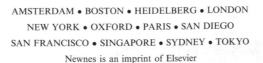

AMSTERDAM • BOSTON • HEIDELBERG • LONDON
NEW YORK • OXFORD • PARIS • SAN DIEGO
SAN FRANCISCO • SINGAPORE • SYDNEY • TOKYO

Newnes is an imprint of Elsevier

ELSEVIER

Newnes

Newnes is an imprint of Elsevier
Linacre House, Jordan Hill, Oxford OX2 8DP, UK
30 Corporate Drive, Suite 400, Burlington, MA 01803, USA

First edition 2002
Reprinted 2003 (twice), 2005
Second edition 2006
Reprinted 2007

Copyright © 2006, Dave Smith. Published by Elsevier Ltd. All rights reserved

The right of Dave Smith to be identified as the author of this work has been
asserted in accordance with the Copyright, Designs and Patents Act 1988

No part of this publication may be reproduced, stored in a retrieval system
or transmitted in any form or by any means electronic, mechanical, photocopying,
recording or otherwise without the prior written permission of the publisher

Permissions may be sought directly from Elsevier's Science & Technology Rights
Department in Oxford, UK: phone (+44) (0) 1865 843830; fax (+44) (0) 1865 853333;
email: permissions@elsevier.com. Alternatively you can submit your request online by
visiting the Elsevier web site at http://elsevier.com/locate/permissions, and selecting
Obtaining permission to use Elsevier material

Notice
No responsibility is assumed by the publisher for any injury and/or damage to persons
or property as a matter of products liability, negligence or otherwise, or from any use
or operation of any methods, products, instructions or ideas contained in the material
herein. Because of rapid advances in the medical sciences, in particular, independent
verification of diagnoses and drug dosages should be made

British Library Cataloguing in Publication Data
A catalogue record for this book is available from the British Library

Library of Congress Cataloging-in-Publication Data
A catalog record for this book is available from the Library of Congress

ISBN–13: 978-0-7506-6826-2
ISBN–10: 0-7506-6826-1

For information on all Newnes publications
visit our website at www.newnespress.com

Printed and bound in *Great Britain*

07 08 09 10 10 9 8 7 6 5 4 3 2

Working together to grow
libraries in developing countries

www.elsevier.com | www.bookaid.org | www.sabre.org

ELSEVIER   BOOK AID
             International   Sabre Foundation

Introduction

The microcontroller is an exciting new device in the field of electronics control. It is a complete computer control system on a single chip. microcontrollers include EPROM program memory, user RAM for storing program data, timer circuits, an instruction set, special function registers, power on reset, interrupts, low power consumption and a security bit for software protection. Some microcontrollers like the 16F818/9 devices include on board A to D converters.

The microcontroller is used as a single chip control unit for example in a washing machine, the inputs to the controller would be from a door catch, water level switch, temperature sensor. The outputs would then be fed to a water inlet valve, heater, motor and pump. The controller would monitor the inputs and decide which outputs to switch on i.e. close the door – water inlet valve open – monitor water level, close valve when water level reached. Check temperature, turn on heater, switch off heater when the correct temperature is reached. Turn the motor slowly clockwise for 5 seconds, anticlockwise for 5 seconds, repeat 20 times, etc. If you are not that maternal maybe you prefer discos to washing – then you can build your own disco lights.

The microcontroller because of its versatility, ease of use and cost will change the way electronic circuits are designed and will now enable projects to be designed which previously were too complex. Additional components such as versatile interface adapters (VIA), RAM, ROM, EPROM and address decoders are no longer required.

One of the most difficult hurdles to overcome when using any new technology is the first one – getting started! It was my aim when writing this book to explain as simply as possible how to program and use the PIC microcontrollers. I hope I have succeeded.

Code examples in this book are available to download from:
http://books.elsevier.com/uk//newnes/uk/subindex.asp?maintarget=companions/
defaultindividual.asp&isbn=0750648120

Dave Smith, B.Sc., M.Sc.
Senior Lecturer in Electronics
Manchester Metropolitan University

1
Introduction to the PIC microcontroller

A microcontroller is a computer control system on a single chip. It has many electronic circuits built into it, which can decode written instructions and convert them to electrical signals. The microcontroller will then step through these instructions and execute them one by one. As an example of this a microcontroller could be instructed to measure the temperature of a room and turn on a heater if it goes cold.

Microcontrollers are now changing electronic designs. Instead of hard wiring a number of logic gates together to perform some function we now use instructions to wire the gates electronically. The list of these instructions given to the microcontroller is called a program.

The aim of the book

The aim of the book is to teach you how to build control circuits using devices such as switches, keypads, analogue sensors, LEDs, buzzers, 7 segment displays, alpha-numeric displays, radio transmitters etc. This is done by introducing graded examples, starting off with only a few instructions and gradually increasing the number of instructions as the complexity of the examples increases.

Each chapter clearly identifies the new instructions added to your vocabulary.

The programs use building blocks of code that can be reused in many different program applications.

Complete programs are provided so that an application can be seen working. The reader is then encouraged to modify the code to alter the program in order to enhance their understanding.

Throughout this book the programs are written in a language called assembly language which uses a vocabulary of 35 words called an instruction set. In order to write a program we need to understand what these words mean and how we can combine them.

The complete instruction set is shown in Chapter 19 Instruction Set, Files and Registers.

All of the programs illustrated in the book are available from:
http://books.elsevier.com/uk//newnes/uk/subindex.asp?maintarget=
companions/defaultindividual.asp&isbn=0750648120

You will of course need a programmer to program the instructions into the chip. The assembler software, MPASM, which converts your text to the machine code is available from Microchip on www.microchip.com this website is a must for PIC programmers.

Program memory

Inside the microcontroller the program we write is stored in an area called **EPROM** (Electrically Programmable Read Only Memory), this memory is non-volatile and is remembered when the power is switched off. The memory is electrically programmed by a piece of hardware called a programmer.

The instructions we program into our microcontroller work by moving and manipulating data in memory locations known as user files and registers. This memory is called **RAM**, Random Access Memory. For example in the room heater we would measure the room temperature by instructing the microcontroller via its Analogue to Digital Control Register (ADCON0) the measurement would then be compared with our data stored in one of the user files. A STATUS Register would indicate if the temperature was above or below the required value and a PORT Register would turn the heater on or off accordingly. The memory map of the 16F84 chip is shown in Chapter 6.

PIC Microcontrollers are 8 bit micros, which means that the memory locations, the user files and registers are made up of 8 binary digits shown in Figure 1.1.

Bit 0 is the Least Significant Bit (LSB) and Bit 7 is the Most Significant Bit (MSB).

bit 7	bit 6	bit 5	bit 4	bit 3	bit 2	bit 1	bit 0
1	0	1	1	0	0	1	0

MSB -- LSB

Figure 1.1 User file and register layout

The use of these binary digits is explained in Appendix C.

When you make an analogue measurement, the digital number, which results, will be stored in a register called ADRES. If you are counting the number of times a light has been turned on and off, the result would be stored as an 8 bit binary number in a user file called, say, COUNT.

Microcontroller clock

In order to step through the instructions the microcontroller needs a clock frequency to orchestrate the movement of the data around its electronic circuits. This can be provided by 2 capacitors and a crystal or by an internal oscillator circuit.

In the 16F84 microcontroller there are 4 oscillator options.

- An RC (Resistor/Capacitor) oscillator which provides a low cost solution.
- An LP oscillator, i.e. 32kHz crystal, which minimises power consumption.
- XT which uses a standard crystal configuration.
- HS is the high-speed oscillator option.

Common crystal frequencies would be 32kHz, 1MHz, 4MHz, 10MHz and 20MHz.

Newer microcontrollers, such as the 16F818 and 12F629, have an oscillator built on the chip so we do not need to add a crystal to them.

Inside the Microcontroller there is an area where the processing (the clever work), such as mathematical and logical operations are performed, this is known as the central processing unit or CPU. There is also a region where event timing is performed and another for interfacing to the outside world through ports.

The microcontroller system

The block diagram of the microcontroller system is shown in Figure 1.2.

Figure 1.2 The basic microcontroller system

- The input components would consist of digital devices such as, switches, push buttons, pressure mats, float switches, keypads, radio receivers etc. and analogue sensors such as light dependant resistors, thermistors, gas sensors, pressure sensors, etc.
- The control unit is of course the microcontroller. The microcontroller will monitor the inputs and as a result the program would turn outputs on and off. The microcontroller stores the program in its memory, and executes the instructions under the control of the clock circuit.
- The output devices would be made up from LEDs, buzzers, motors, alpha numeric displays, radio transmitters, 7 segment displays, heaters, fans etc.

The most obvious choice then for the microcontroller is how many digital inputs, analogue inputs and outputs does the system require. This would then specify the minimum number of inputs and outputs (I/O) that the microcontroller must have. If analogue inputs are used then the microcontroller must have an Analogue to Digital (A/D) module inside.

The next consideration would be what size of program memory storage is required. This should not be too much of a problem when starting out, as most programs would be relatively small. All programs in this book fit into a 1k program memory space.

The clock frequency determines the speed at which the instructions are executed. This is important if any lengthy calculations arc bcing undertaken. The higher the clock frequency the quicker the micro will finish one task and start another.

Other considerations are the number of interrupts and timer circuits required, how much data EEPROM if any is needed. These more complex operations are considered later in the text.

In this book the programs requiring analogue inputs have been implemented on the 16F818 and 16F872 micros. Programs requiring only digital inputs have used the 16F84 and 16F818. The 16F818 and 16F84 devices have 1k of program memory and have been run using a 32.768kHz clock frequency or the internal oscillator on the 16F818. There are over 100 PIC microcontrollers, the problem of which one to use need not be considered until you have understood a few applications.

Types of microcontroller

The list of PIC Microcontrollers is growing almost daily. They include devices for all kinds of applications, for example the 18F8722 has 64k of EPROM memory, 3938 bytes of RAM (User files), 1024 bytes of EEPROM, 16 10-bit

A/D channels, a voltage reference, 72 inputs and outputs (I/O), 3–16 bit and 2–8 bit timers.

There are basically two types of microcontrollers, Flash devices and One Time Programmable Devices (OTP).

The flash devices can be reprogrammed in the programmer whereas OTP devices once programmed cannot be reprogrammed. All OTP devices however do have a windowed variety, which enables them to be erased under ultra violet light in about 15 minutes, so that they can be reprogrammed. The windowed devices have a suffix JW to distinguish them from the others.

The OTP devices are specified for a particular oscillator configuration R-C, LP, XT or HS. See Appendix A Microcontroller Data.

16C54 configurations are:

16C54JW	Windowed device
16C54RC	OTP, R-C oscillator
16C54LP	OTP, LP oscillator, 32kHz
16C54XT	OTP, XT oscillator, 4MHz
16C54HS	OTP, HS oscillator, 20Mhz

In this book the two main devices investigated are the 16F84 and the 16F818 flash devices. The 16F84 at present is the main choice for beginners, but should be replaced in popularity by the better and cheaper 16F818. They have their program memory made using Flash technology. They can be programmed, tested in a circuit and reprogrammed if required without the need for an ultra violet eraser.

Microcontroller specification

You specify a device with its Product Identification Code.
This code specifies:

- The device number.
- If it is a Windowed, an OTP, or flash device. The windowed device is specified by a JW suffix. OTP devices are specified by Oscillator Frequency, and the Flash devices are specified with an F such as 16F84.
- The oscillation frequency, usually 04 for devices working up to 4MHz., 10 up to 10MHz or 20 up to 20MHz. 20MHz devices are of course more expensive than 4MHz devices.
- Temperature range, for general applications 0°C to +70°C is usually specified.

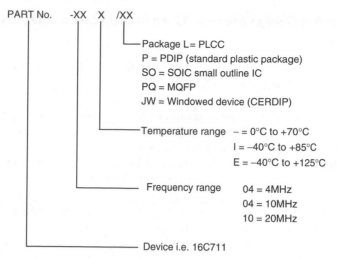

Figure 1.3 Product identification system

The Product Identification System for the PIC Micro is shown in Figure 1.3.

Using the microcontroller

In order to use the microcontroller in a circuit there are basically two areas you need to understand:

1. How to connect the microcontroller to the hardware.
2. How to write and program the code into the microcontroller.

1 Microcontroller hardware

The hardware that the microcontroller needs to function is shown in Figure 1.4. The crystal and capacitors connected to pins 15 and 16 of the 16F84 produce the clock pulses that are required to step the microcontroller through the program and provide the timing pulses. (The crystal and capacitor can be omitted if using an on board oscillator in e.g. 16F818). The 0.1μF capacitor is placed as close to the chip as possible between 5v and 0v. Its role is to divert (filter) any electrical noise on the 5v power supply line to 0v, thus bypassing the microcontroller. This capacitor must always be connected to stop any noise affecting the normal running of the microcontroller.

Microcontroller power supply

The power supply for the microcontroller needs to be between 2v and 6v. This can easily be provided from a 6v battery as shown in Figure 1.5.

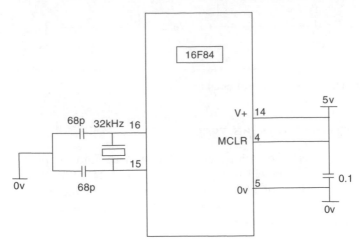

Figure 1.4 The microcontroller circuit

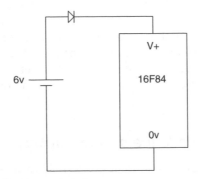

Figure 1.5 Microcontroller power supply

The diode in the circuit drops 0.7v across it reducing the applied voltage to 5.3v. It provides protection for the microcontroller if the battery is accidentally connected the wrong way round. In that case the diode would be reversed biased and no current would flow.

7805, Voltage regulator circuit

Probably the most common power supply connection for the microcontroller is a 3 terminal voltage regulator, I.C., the 7805. The connection for this is shown in Figure 1.6.

The supply voltage, Vin, to the 7805 can be anything from 7v to 30v.

The output voltage will be a fixed 5v and can supply currents up to 1amp. So battery supplies such as 24v, 12v, 9v etc. can be accommodated.

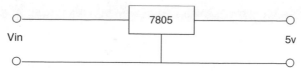

Figure 1.6 The voltage regulator circuit

Power dissipation in the 7805

Care must be taken when using a high value for Vin. For example if Vin $= 24$v the output of the 7805 will be 5v, so the 7805 has $24 - 5 = 19$v across it. If it is supplying a current of 0.5amp to the circuit then the power dissipated (volts $\times$ current) is $19 \times 0.5 = 9.5$watts. The regulator will get hot! and will need a heat sink to dissipate this heat.

If a supply of 9v is connected to the regulator it will have 4v across it and would dissipate $4 \times 0.5 = 2$watts.

In the circuits used in this book the microcontroller only requires a current of 15µA so most of the current drawn will be from the outputs. If the output current is not too large say <100mA (0.1A) then with a 9v supply the power dissipated would be $4 \times 0.1 = 0.4$watts and the regulator will stay cool without a heatsink.

Connecting switches to the microcontroller

The most common way of connecting a switch to a microcontroller is via a pull-up resistor to 5v as shown in Figure 1.7.

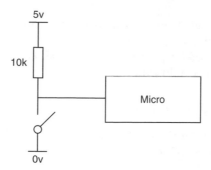

Figure 1.7 Connecting a switch to the microcontroller

When the switch is open, 5v, a logic 1 is connected to the micro.

When the switch is closed, 0v, a logic 0 is connected to the micro.

Some Microcontrollers such as the 16F84 and 16F818 have internal pull ups connected to some of their I/O pins. PORTB in the above devices.

Figure 1.8 shows how the switch is connected using the internal pull up.

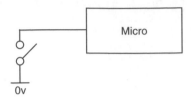

Figure 1.8 Connecting a switch using an internal pull up

Connecting outputs to the microcontroller

The microcontroller is capable of supplying approximately 20–25mA to an output pin. So loads such as LEDs or small relays can be driven directly. Larger loads require interfacing via a transistor, for dc or a triac, for ac. Opto-coupled devices provide an isolated interface between the microcontroller and the load.

The LED connection to the Micro is shown in Figure 1.9.

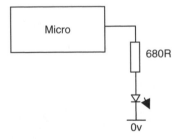

Figure 1.9 Connecting an LED to the microcontroller

2 Programming the microcontroller

In order to have the microcontroller perform some controlling action you need to communicate with it and tell it what those instructions are to be. When we communicate with one another we use a spoken language, when we communicate with a microcontroller we use a program language. The program language for the PIC Microcontroller uses 35 words (instructions)

in its vocabulary. A few more instructions are used in the bigger microcontrollers.

In order to communicate with the microcontroller we need to know what these 35 instructions are and how to use them. Not all 35 instructions are used in this book. In fact you can write meaningful programs using only 5 or 6 instructions.

2
Programming the 16F84 microcontroller

Microcontrollers are now providing us with a new way of designing circuits. Designs, which at one time required many Digital ICs and lengthy Boolean Algebra calculations, can now be programmed simply into one Micro-controller. For example a set of traffic lights would have required an oscillator circuit, counting and decoding circuits plus an assortment of logic gate ICs.

In order to use this exciting new technology we must learn how to program these Microcontrollers.

The Microcontroller I have chosen to start with is the 16F84-04/P, which means it is a flash device that can be electrically erased and reprogrammed without using an Ultra Violet Eraser. It can be used up to an oscillation frequency of 4MHz and comes in a standard 18pin Plastic package.

It has 35 instructions in its vocabulary, but like all languages not all of the instructions are used all of the time you can go a long way on just a few. In order to teach you how to use these instructions I have started off with a simple program to flash an LED on and off continually. This program introduces you to 4 instructions in 5 lines of code.

You are then encouraged to write your own program to flash two LEDs on and off alternately. The idea being, when you have understood my code you can then modify it for your own program, thus understanding better. Once you have written your first program you are then off and running. The book then continues with further applications such as traffic lights and disco lights to introduce more of the instructions increasing your microcontroller vocabulary.

Instructions used in this chapter:

- BCF
- BSF
- CALL
- GOTO

Microcontroller inputs and outputs (I/O)

The microcontroller is a very versatile chip and can be programmed to operate in a number of different configurations. The 16F84 is a 13 I/O device, which means it has 13 Inputs and Outputs. The I/O can be configured in any combination i.e. 1 input 12 outputs, 6 inputs 7 outputs, or 13 outputs depending on your application. These I/O are connected to the outside world through registers called Ports. The 16F84 has two ports, PORTA and PORTB. PORTA is a 5-bit port it has 5 I/O lines and PORTB has 8 I/O.

Timing with the microcontroller

All microcontrollers have timer circuits onboard; some have 4 different timers. The 16F84 has one timer register called TIMER0. These timers run at a speed of ¼ of the clock speed. So if we use a 32,768Hz crystal the internal timer will run at ¼ of 32768Hz i.e. 8192Hz. If we want to turn an LED on for say 1 second we would need to count 8192 of these timing pulses. This is a lot of pulses! Fortunately within the microcontroller there is a register called an OPTION Register, that allows us to slow down these pulses by a factor of 2, 4, 8, 16, 32, 64, 128 or 256. The OPTION Register is discussed in the Instruction Set, Files and Register section in Chapter 19. Setting the prescaler, as it is called to divide by 256 in the OPTION register means that our timing pulses are now 8192/256 = 32Hz, i.e. 32 pulses a second. So to turn our LED on for 1 second we need only to count 32 pulses in TIMER0, or 16 for 0.5 seconds, or 160 for 5 seconds etc.

Programming the microcontroller

In order to program the microcontroller we need to:

• Write the instructions in a program.
• Change the text into machine code that the microcontroller understands using a piece of software called an assembler.
• Blow the data into the chip using a programmer.

Let's consider the first task, writing the program. This can be done on any text editor, such as notepad. I prefer to use an editor supplied by the microcontroller manufacturers, 'Microchip'. This software is called MPLAB and is available free on www.microchip.com.

As you have seen above we need to configure the I/O and set the Prescaler for the timing. If we do not set them the default conditions are that all **PORT** bits are inputs. A micro with no outputs is not much use! The default for the Prescaler is that the clock rate is divided by 2.

The program also needs to know what device it is intended for and also what the start address in the memory is.

If this is starting to sound confusing – do not worry, I have written a header program, which sets the all the above conditions for you to use. These conditions can be changed later when you understand more about what you are doing.

The header for the 16F84 sets the 5 bits of PORTA as inputs and the 8 bits of PORTB as outputs. It also sets the prescaler to divide by 256. We will use the 32,768Hz crystal so our timing is 32 pulses per sec. The program instructions will run at ¼ of the 32,768Hz clock, i.e. 8192 instructions per second. The header also includes two timing subroutines for you to use they are DELAY1 – a 1 second delay and DELAYP5 – a half-second delay. A subroutine is a section of code that can be called, when needed, to save writing it again. For the moment do not worry about how the header or the delay subroutines work. We will work through them, in Chapter 6, once we have programmed a couple of applications.

Just one more point, the different ways of entering data.

Entering data

Consider the decimal number 37, this has a Hex value of 25 or a Binary value of 0010 0101. The assembler will accept this as .37 in decimal (note the . is not a decimal point) or as 25H in hex or B'00100101' in binary.

181 decimal would be entered as .181 in decimal, 0B5H in hex or B'10110101' in binary. NB. If a hex number starts with a letter it must be prefixed with a 0, i.e. 0B5H not B5H.

NB. The default radix for the assembler MPASM is hex.

Appendix C. illustrates how to change between Decimal, Binary and Hexadecimal numbers.

The PIC Microcontrollers are 8 bit micros. This means that the memory locations, i.e. user files and registers contain 8 bits. So the smallest 8 bit number is of course 0000 0000 which is equal to a decimal number 0 (of course). The largest 8 bit number is 1111 1111 which is equal to a decimal number of 255. To use numbers bigger than 255 we have to combine memory locations. Two memory locations combine to give 16 bits with numbers up to 65,536. Three memory locations combine to give 24 bits allowing numbers up to 16,777,215

and so on. These large numbers are introduced in Chapter 8, Numbers Larger than 255.

The Header for the 16F84, HEADER84.ASM

The listing below shows the header for the 16F84 microcontroller. I suggest you start all of your programs, for this chip, with this header, or a modified version of it. A full explanation of this header file is given in Chapter 6.

; HEADER84.ASM for 16F84. This sets PORTA as an INPUT (NB 1means input).
; and PORTB as an OUTPUT (NB 0 means output).
;The OPTION Register is set to /256 to give timing pulses of 1/32 of a second.
;1second and 0.5 second delays are included in the subroutine section.

;***

; EQUATES SECTION

```
TMR0       EQU    1       ;means TMR0 is file 1.
STATUS     EQU    3       ;means STATUS is file 3.
PORTA      EQU    5       ;means PORTA is file 5.
PORTB      EQU    6       ;means PORTB is file 6.
TRISA      EQU    85H     ;TRISA (the PORTA I/O selection) is file 85H
TRISB      EQU    86H     ;TRISB (the PORTB I/O selection) is file 86H
OPTION_R   EQU    81H     ;the OPTION register is file 81H
ZEROBIT    EQU    2       ;means ZEROBIT is bit 2.
COUNT      EQU    0CH     ;COUNT is file 0C, a register to count events.
```

;***

```
LIST       P = 16F84     ; we are using the 16F84.
ORG        0             ;the start address in memory is 0
GOTO       START         ; goto start!
```

;***

; Configuration Bits
__CONFIG H'3FF0'   ;selects LP oscillator, WDT off, PUT on,
                    ;Code Protection disabled.

;***

;SUBROUTINE SECTION.

;1 second delay.

Contents

```
DELAY1      CLRF        TMR0            ;START TMR0.
LOOPA       MOVF        TMR0,W          ;READ TMR0 INTO W.
            SUBLW       .32             ;TIME - 32
            BTFSS       STATUS,
                        ZEROBIT         ; Check TIME-W = 0
            GOTO        LOOPA           ;Time is not = 32.
            RETLW       0               ;Time is 32, return.

; 0.5 second delay.
DELAYP5     CLRF        TMR0            ;START TMR0.
LOOPB       MOVF        TMR0,W          ;READ TMR0 INTO W.
            SUBLW       .16             ;TIME - 16
            BTFSS       STATUS,
                        ZEROBIT         ; Check TIME-W = 0
            GOTO        LOOPB           ;Time is not = 16.
            RETLW       0               ;Time is 16, return.
;*********************************************************

;CONFIGURATION SECTION

START       BSF         STATUS,5        ;Turns to Bank1.
            MOVLW       B'00011111'     ;5bits of PORTA are I/P
            MOVWF       TRISA

            MOVLW       B'00000000'
            MOVWF       TRISB           ;PORTB is OUTPUT

            MOVLW       B'00000111'     ;Prescaler is /256
            MOVWF       OPTION_R        ;TIMER is 1/32 secs.
            BCF         STATUS,5        ;Return to Bank0.
            CLRF        PORTA           ;Clears PortA.
            CLRF        PORTB           ;Clears PortB.
;*********************************************************
;Program starts now.

END         ;This must always come at the end of your code
```

NB. In the program any text on a line following the semicolon (;) is ignored by the assembler software. Program comments can then be placed there.

The section is saved as HEADER84.ASM you can use it to start all of your 16F84 programs. HEADER84 is the name of our program and ASM is its extension.

Program example

The best way to begin to understand how to use a microcontroller is to start with a simple example and then build on this.

Let us consider a program to flash an LED ON and OFF at 0.5 second intervals. The LED is connected to PortB bit 0 as shown in Figure 2.1.

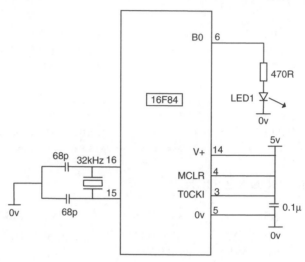

Figure 2.1 Circuit diagram of the microcontroller flasher

Notice from Figure 2.1 how few components the microcontroller needs – 2 × 68pF capacitors, a 32.768kHz crystal for the oscillator and a 0.1μF capacitor for decoupling the power supply. Other oscillator and crystal configurations are possible – see Microchip's data sheets for other combinations. I have chosen the 32kHz crystal because it enables times of seconds to be produced easily.

The program for this circuit can be written on any text editor, such as Notepad or on Microchip's editor MPLAB.

Open HEADER84.ASM or start a new file and type the program in, saving as HEADER84.ASM If using Notepad saveas type "All Files" to avoid Notepad adding the extension .TXT

Once you have HEADER84.ASM saved on disk and loaded onto the screen alter it by including your program as shown below:-

; HEADER84.ASM for 16F84. This sets PORTA as an INPUT (NB 1means input).

; and **PORTB** as an OUTPUT (NB 0 means output).
;The OPTION Register is set to /256 to give timing pulses of 1/32 of a second.
;1second and 0.5 second delays are included in the subroutine section.

;**

; EQUATES SECTION

TMR0	EQU	1	;means TMR0 is file 1.
STATUS	EQU	3	;means STATUS is file 3.
PORTA	EQU	5	;means PORTA is file 5.
PORTB	EQU	6	;means PORTB is file 6.
TRISA	EQU	85H	;TRISA (the PORTA I/O selection) is file 85H
TRISB	EQU	86H	;TRISB (the PORTB I/O selection) is file 86H
OPTION_R	EQU	81H	;the OPTION register is file 81H
ZEROBIT	EQU	2	;means ZEROBIT is bit 2.
COUNT	EQU	0CH	; COUNT is file 0C, a register to count events.

;**

LIST	P = 16F84	;we are using the 16F84.
ORG	0	;the start address in memory is 0
GOTO	START	;goto start!

;**
; Configuration Bits

__CONFIG H'3FF0'   ;selects LP oscillator, WDT off, PUT on,
                          ;Code Protection disabled.

;**

;SUBROUTINE SECTION.

; 1 second delay.

DELAY1	CLRF	TMR0	;START TMR0.
LOOPA	MOVF	TMR0,W	;READ TMR0 INTO W.
	SUBLW	.32	;TIME - 32
	BTFSS	STATUS, ZEROBIT	; Check TIME-W = 0
	GOTO	LOOPA	;Time is not = 32.
	RETLW	0	;Time is 32, return.

; 0.5 second delay.

```
DELAYP5   CLRF      TMR0         ;START TMR0.
LOOPB     MOVF      TMR0,W       ;READ TMR0 INTO W.
          SUBLW     .16          ;TIME - 16
          BTFSS     STATUS,
                    ZEROBIT      ; Check TIME-W = 0
          GOTO      LOOPB        ;Time is not = 16.
          RETLW     0            ;Time is 16, return.
```

;***

;CONFIGURATION SECTION.

```
START     BSF       STATUS,5     ;Turns to Bank1.

          MOVLW     B'00011111'  ;5bits of PORTA are I/P
          MOVWF     TRISA

          MOVLW     B'00000000'
          MOVWF     TRISB        ;PORTB is OUTPUT
          MOVLW     B'00000111'  ;Prescaler is /256
          MOVWF     OPTION_R     ;TIMER is 1/32 secs.

          BCF       STATUS,5     ;Return to Bank0.
          CLRF      PORTA        ;Clears PortA.
          CLRF      PORTB        ;Clears PortB.
```
;***
;Program starts now.

```
BEGIN     BSF       PORTB,0      ;Turn ON B0.
          CALL      DELAYP5      ;Wait 0.5 seconds
          BCF       PORTB,0      ;Turn OFF B0.
          CALL      DELAYP5      ;Wait 0.5 seconds
          GOTO      BEGIN        ;Repeat
END                              ;YOU MUST END!!
```

How Does It Work?

The 5 lines of code starting at BEGIN are responsible for flashing the LED ON and OFF. This is all the code we will require for now. The rest of the code, the header is explained in Chapter 6 once you have seen the program working.

- BEGIN is a label. A label is used as a location for the program to go to.
- Line1 the instruction BSF and its data PORTB,0 is shorthand for Bit Set in File, which means Set the Bit in the File PORTB, where bit0 is the designated bit. This will cause PORTB,0 to be Set to a logic1, in hardware terms this means pin6 in Figure 2.1 is at 5v turning the LED on.

NB. There must not be any spaces in a label, an instruction or its data. I keep the program tidy by using the TAB key on the keyboard.

- Line2 CALL DELAYP5 causes the program to wait 0.5 seconds while the subroutine DELAYP5 in the header is executed.
- Line3 BCF PORTB,0 is the opposite of Line1, this code is shorthand for Bit Clear in File, which means Clear the Bit in the File PORTB, where bit0 is the designated bit. This will cause PORTB,0 to be Cleared to a logic0, in hardware terms this means pin6 in Figure 2.1 is at 0v turning the LED off.
- Line4 CALL DELAYP5 is the same as Line2.
- Line5 GOTO BEGIN sends the program back to the label BEGIN to repeat the process of flashing the LED on and off.

Any of the 8 outputs can be turned ON and OFF using the 2 instructions BSF and BCF for example:

BSF PORTB,3 makes PORTB,3 (pin9) 5v.
BCF PORTB,7 makes PORTB,7 (pin13) 0v.

Saving and assembling the code

The program is then saved as FLASHER.ASM. The next task is to assemble this text into the HEX code that the microcontroller understands.

Open MPLAB the screen shown below in Figure 2.2 will open up.

Open the file FLASHER.ASM using the FILE menu as shown in Figure 2.3.

From the CONFIGURE Menu, Select Device then choose the micro 16F84 in this example, as indicated in Figure 2.4.

Next choose CONFIGURE – Configuration Bits as shown in Figure 2.5 and set as indicated.

Our configuration bits setting, select the LP Oscillator, turn the Watchdog Timer Off, turn the Power Up Timer on and turn Code Protect off.

Notice the value of this configuration is 3FF0 in hex. This configuration setting can be written into the header program so there is no need to here. The code is __CONFIG H'3FF0'

The choice of configuration bit settings for the 16F84 are:

- the Oscillator, RC, LP, XT, HS. i.e. LP
- Watchdog Timer ON/OFF i.e. OFF

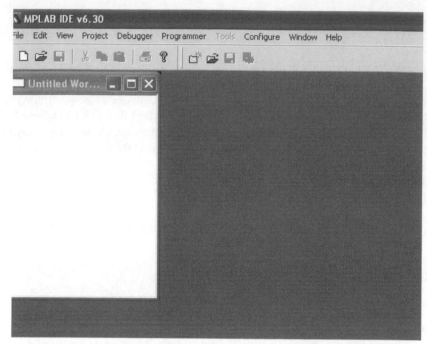

Figure 2.2   MPLAB initial screen

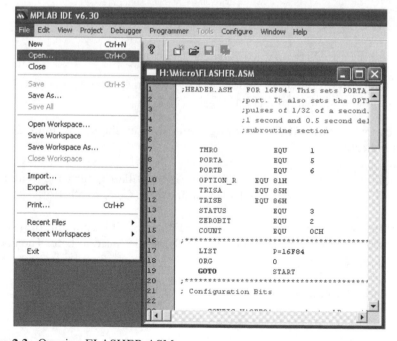

Figure 2.3   Opening FLASHER.ASM

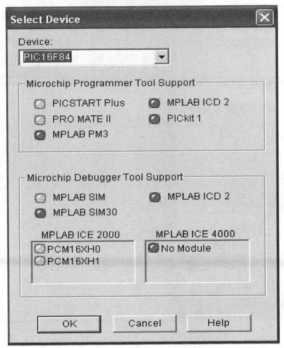

Figure 2.4 CONFIGURE – select device

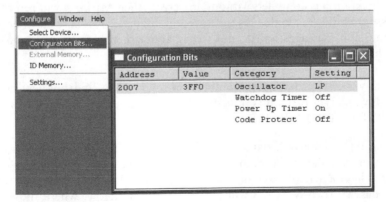

Figure 2.5 Configuration bits setting

- Power Up Timer ON/OFF i.e. ON
- Code Protect ON/OFF i.e. OFF

Then we have to convert our text, FLASHER.ASM into a machine code file FLASHER.HEX to do this choose PROJECT – Quickbuild Flasher.ASM as shown in Figure 2.6.

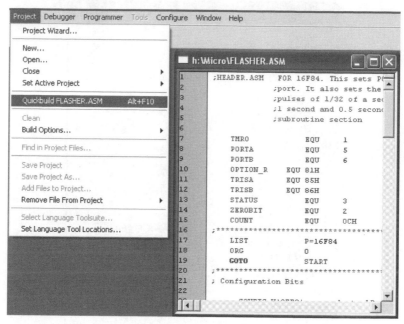

Figure 2.6   Compiling FLASHER.ASM to FLASHER.HEX

If the program has compiled without any errors then MPLAB will return with a message Build Succeeded as indicated in Figure 2.7. There may be some warnings and messages but do not worry about them, the compiler has seen something it wasn't expecting.

Incidently, I always have line numbers on my code to find my way around, especially in larger programs. Line numbers can be turned on and off with the path: EDIT – PROPERTIES.

Suppose that you have a syntax error in your code. The message Build Failed will appear as shown in Figure 2.8. You then have to correct the errors. MPLAB has indicated the error in the message box. If you 'double click' on the error message then MPLAB will indicate, with an arrow, where the error is

Figure 2.7 Build Succeeded

in your code. Correct the errors and compile (Quickbuild) again to produce an error free build.

The error I have written into my code occurs in line 61, with the message, 'symbol not previously defined (PORT)'. I should have written PORTB the compiler does not understand 'PORT'.

After successfully building the program, the HEX code is ready to be programmed into the Microcontroller.

You can view your compilation using VIEW – PROGRAM MEMORY as shown in Figure 2.9.

The FLASHER.HEX file is now ready to be programmed into the chip.

PICSTART PLUS programmer

If you do not have a programmer I would recommend Arizona Microchip's own PICSTART PLUS. When Arizona bring out a new microcontroller as

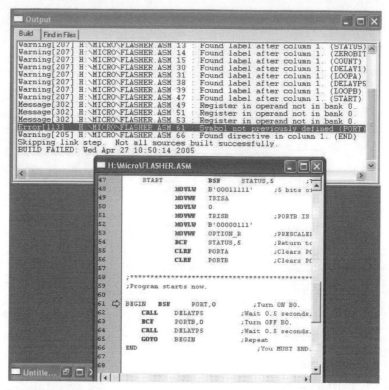

Figure 2.8 Build failed

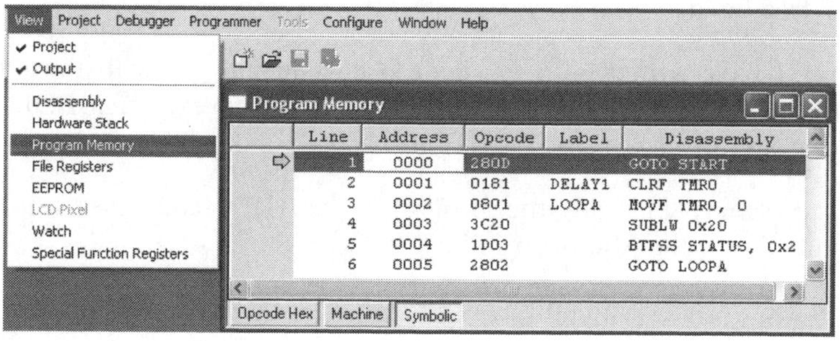

Figure 2.9 Program memory

they do regularly, the driver software is updated and can be downloaded free off the internet from MICROCHIP.COM.

Once installed on your PC it is opened from MPLAB i.e.

Switch on the PICSTART Plus Programmer.

Select, Programmer – Select Programmer – PICSTART Plus, shown in Figure 2.10.

Figure 2.10 Selecting the PICSTART plus programmer

Select Enable Programmer from the Programmer box, Figure 2.10.

The final stage is to program your code onto the chip. To do this click the programming icon shown in Figure 2.11 or via the menu on Programmer – Program.

Figure 2.11 Programming icon

After a short while the message success will appear on the screen.

You will be greeted with the success statement for a few seconds only, if you miss it check the program statistics for Pass 1 Fail 0 Total 1, which will be continually updated.

The code has been successfully blown into your chip and is ready for use.

If this process fails – check the chip is inserted correctly in the socket, if it is then try another chip.

So we are now able to use the microcontroller to switch an LED on and off – Fantastic!

But use your imagination. There are 35 instructions in your micro vocabulary. The PIC Microcontroller range at the moment includes devices with 64k bytes of EPROM-program memory, 3938 bytes of RAM-data memory, 1024 bytes of EEPROM, 72 Input and Output pins, 11 interrupts, 15 channel A/D converter, 20MHz. clock, real time clock, 4 counter/timers, 55 word instruction set. See Appendix A for a detailed list. If the 64k of EPROM or 3938 bytes of RAM is not enough your system can be expanded using extra EPROM and RAM. In the end the only real limits will be your imagination.

Programming flowchart

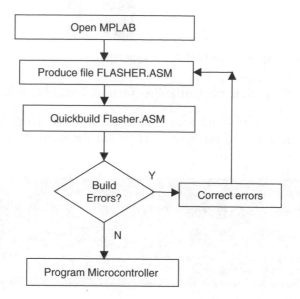

Problem: flashing two LEDs

There has been a lot to do and think about to get this first program into the microcontroller and make it work in a circuit. But just so that you are sure what you are doing – Write a program that will flash two LEDs on and off alternately. Put LED0 on B0 and LED1 on B1. NB you can use the file FLASHER.ASM it only needs two extra lines adding! Then save it as FLASHER2.ASM

The circuit layout is shown in Figure 2.12.

Try not to look at the solution below before you have attempted it.

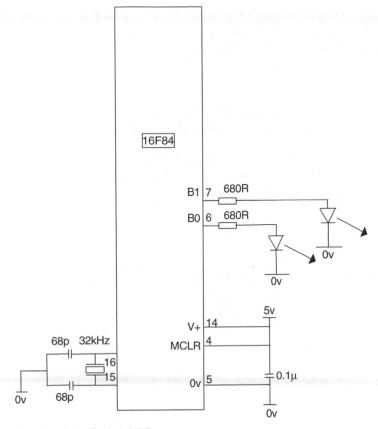

Figure 2.12 Circuit to flash 2 LEDs

Solution to the problem, flashing two LEDs

The header is the same as in FLASHER.ASM. just include in the section, program starts now, the following lines:

;Program starts now.

```
BEGIN     BSF        PORTB,0      ;Turn ON B0.
          BCF        PORTB,1      ;Turn OFF B1
          CALL       DELAYP5      ;Wait 0.5 seconds
          BCF        PORTB,0      ;Turn OFF B0.
          BSF        PORTB,1      ;Turn ON B1.
          CALL       DELAYP5      ;Wait 0.5 seconds
          GOTO       BEGIN        ;Repeat
END
```

Did you manage to do this? If not have a look at my solution and see what the lines are doing. Now try flashing 4 LEDs on and off, with 2 on and two off alternately. You might like to have them on for 1 second and off for half a second. Can you see how to use the 1-second delay in place of the half-second delay.

The different combinations of switching any 8 LEDs on PORTB should be relatively easy once you have mastered these steps.

Perhaps the most difficult step in understanding any new technology is getting started. The next chapter will introduce a few more projects similar to Flasher.ASM to help you progress.

3
Introductory projects

New instructions used in this chapter:

- MOVLW
- MOVWF
- DECFSZ

Let's have a look at a few variations of flashing the LEDs on and off to develop our programming skills.

LED_Flasher2

Suppose we want to switch the LED on for 2 seconds and off for 1 second. Figure 2.1 shows the circuit diagram for this application. The code for this would be:

```
;Program starts now.
BEGIN       BSF         PORTB,0     ;Turn on B0
            CALL        DELAY1      ;Wait 1 second
            CALL        DELAY1      ;Wait 1 second
            BCF         PORTB,0     ;Turn off B0
            CALL        DELAY1      ;Wait 1 second
            GOTO        BEGIN       ;Repeat
END
```

NB. This code would be added to HEADER84.ASM into the section called, "Program starts now".

To do this open MPLAB, then FILE – OPEN – HEADER84.ASM

Add the code and saveas LED_FLASHER2.ASM

The text would then be assembled by the MPLAB software and then blown into the Microcontroller as explained in Chapter 2.

How does it work?

The comments alongside the code explain what the lines are doing. Because we do not have a 2 second delay we wait for 1 second twice. You can of course write a 2 second delay routine but we will be looking at this later.

SOS

For our next example let us switch B0 on and off just as we have been doing but this time we will use delays of ¼ second and ½ second. This is not much different than we have done previously, but instead of turning an LED on and off we will replace it by a buzzer. The program is not just going to turn a buzzer on and off, but do it in a way that generates the signal, SOS. Which is DOT,DOT,DOT DASH,DASH,DASH DOT,DOT, DOT. Where the DOT is the buzzer on for ¼ second and the DASH is the buzzer on for ½ second with ¼ second between the beeps.

The circuit diagram for the SOS circuit is shown in Figure 3.1.

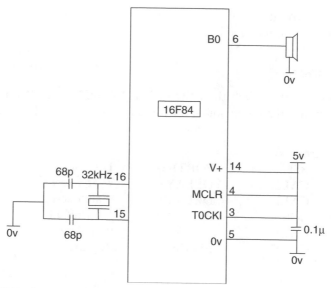

Figure 3.1 SOS circuit diagram

Code for SOS circuit

The complete code for the SOS circuit is shown below because an extra subroutine, DELAYP25, has been added.

```
;SOS.ASM for 16F84. This sets PORTA as an INPUT (NB 1means input)
;and PORTB as an OUTPUT (NB 0 means output).
;The OPTION Register is set to /256 to give timing pulses of 1/32 of a second.
;1second, 0.5 second and 0.25 second delays are included in the subroutine
;section.
```

```
;****************************************************
;EQUATES SECTION

TMR0        EQU    1        ;means TMR0 is file 1.
STATUS      EQU    3        ;means STATUS is file 3.
PORTA       EQU    5        ;means PORTA is file 5.
PORTB       EQU    6        ;means PORTB is file 6.
TRISA       EQU    85H      ;TRISA (the PORTA I/O selection) is file 85H
TRISB       EQU    86H      ;TRISB (the PORTB I/O selection) is file 86H
OPTION_R    EQU    81H      ;the OPTION register is file 81H
ZEROBIT     EQU    2        ;means ZEROBIT is bit 2.
COUNT       EQU    0CH      ;COUNT is file 0C, a register to count events.

;****************************************************

LIST        P=16F84         ;we are using the 16F84.
ORG         0               ;the start address in memory is 0
GOTO        START           ;goto start!

;**********************************************************

;Configuration Bits

    __CONFIG H'3FF0'        ;selects LP oscillator, WDT off, PUT on,
                            ;Code Protection disabled.

;****************************************************
;SUBROUTINE SECTION

;1 second delay.
DELAY1    CLRF      TMR0             ;START TMR0.
LOOPA     MOVF      TMR0,W           ;READ TMR0 INTO W.
          SUBLW     .32              ;TIME - 32
          BTFSS     STATUS,
                    ZEROBIT          ;Check TIME-W=0
          GOTO      LOOPA            ;Time is not = 32.
          RETLW     0                ;Time is 32, return.


;0.5 second delay.
DELAYP5   CLRF      TMR0             ;START TMR0.
LOOPB     MOVF      TMR0,W           ;READ TMR0 INTO W.
          SUBLW     .16              ;TIME - 16
          BTFSS     STATUS,
                    ZEROBIT          ;Check TIME-W=0
          GOTO      LOOPB            ;Time is not = 16.
          RETLW     0                ;Time is 16, return.
```

```
;0.25 second delay.
DELAYP25  CLRF       TMR0           ;START TMR0.
LOOPC     MOVF       TMR0,W         ;READ TMR0 INTO W.
          SUBLW      .8             ;TIME - 8
          BTFSS      STATUS,
                     ZEROBIT        ;Check TIME-W = 0
          GOTO       LOOPC          ;Time is not = 8.
          RETLW      0              ;Time is 8, return.
```

;**
;CONFIGURATION SECTION

```
START     BSF        STATUS,5       ;Turns to Bank1.

          MOVLW      B'00011111'    ;5bits of PORTA are I/P
          MOVWF      TRISA

          MOVLW      B'00000000'
          MOVWF      TRISB          ;PORTB is OUTPUT

          MOVLW      B'00000111'    ;Prescaler is /256
          MOVWF      OPTION_R       ;TIMER is 1/32 secs.

          BCF        STATUS,5       ;Return to Bank0.
          CLRF       PORTA          ;Clears PortA.
          CLRF       PORTB          ;Clears PortB.
```

;**
;Program starts now.

```
BEGIN     BSF        PORTB,0        ;Turn ON B0, DOT
          CALL       DELAYP25       ;Wait 0.25 seconds
          BCF        PORTB,0        ;Turn OFF B0.
          CALL       DELAYP25       ;Wait 0.25 seconds
          BSF        PORTB,0        ;Turn ON B0, DOT
          CALL       DELAYP25       ;Wait 0.25 seconds
          BCF        PORTB,0        ;Turn OFF B0.
          CALL       DELAYP25       ;Wait 0.25 seconds
          BSF        PORTB,0        ;Turn ON B0, DOT
          CALL       DELAYP25       ;Wait 0.25 seconds
          BCF        PORTB,0        ;Turn OFF B0.
          CALL       DELAYP5        ;Wait 0.5 seconds

          BSF        PORTB,0        ;Turn ON B0, DASH
          CALL       DELAYP5        ;Wait 0.5 seconds
```

```
BCF      PORTB,0      ;Turn OFF B0.
CALL     DELAYP25     ;Wait 0.25 seconds
BSF      PORTB,0      ;Turn ON B0, DASH
CALL     DELAYP5      ;Wait 0.5 seconds
BCF      PORTB,0      ;Turn OFF B0.
CALL     DELAYP25     ;Wait 0.25 seconds
BSF      PORTB,0      ;Turn ON B0, DASH
CALL     DELAYP5      ;Wait 0.5 seconds
BCF      PORTB,0      ;Turn OFF B0.
CALL     DELAYP5      ;Wait 0.5 seconds

BSF      PORTB,0      ;Turn ON B0, DOT
CALL     DELAYP25     ;Wait 0.25 seconds
BCF      PORTB,0      ;Turn OFF B0.
CALL     DELAYP25     ;Wait 0.25 seconds
BSF      PORTB,0      ;Turn ON B0, DOT
CALL     DELAYP25     ;Wait 0.25 seconds
BCF      PORTB,0      ;Turn OFF B0.
CALL     DELAYP25     ;Wait 0.25 seconds
BSF      PORTB,0      ;Turn ON B0, DOT
CALL     DELAYP25     ;Wait 0.25 seconds
BCF      PORTB,0      ;Turn OFF B0.
CALL     DELAYP5      ;Wait 0.5 seconds

CALL     DELAY1
CALL     DELAY1       ;Wait 2 seconds before returning.
GOTO     BEGIN        ;Repeat
END                   ;YOU MUST END!!
```

How does it work?

I think the explanation of the code is clear from the comments. At the end of the SOS the program has a delay of 2 seconds before repeating. This should be a useful addition to any alarm project.

We will now move onto switching a number of outputs on and off. Consider flashing all 8 outputs on PORTB on and off at ½ second intervals.

Flashing 8 LEDs

The circuit for this is shown in Figure 3.2.

This code is to be added to HEADER84.ASM as in LED_FLASHER2.ASM

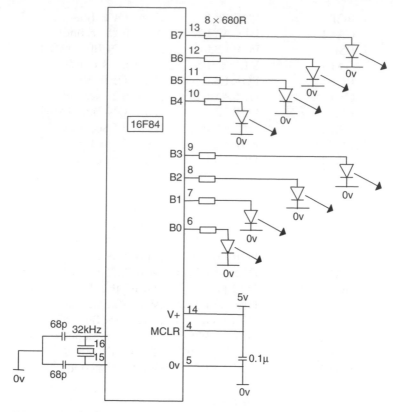

Figure 3.2 Flashing 8 LEDs

;Program starts now.

BEGIN	BSF	PORTB,0	;Turn ON B0
	BSF	PORTB,1	;Turn ON B1
	BSF	PORTB,2	;Turn ON B2
	BSF	PORTB,3	;Turn ON B3
	BSF	PORTB,4	;Turn ON B4
	BSF	PORTB,5	;Turn ON B5
	BSF	PORTB,6	;Turn ON B6
	BSF	PORTB,7	;Turn ON B7
	CALL	DELAYP5	;Wait 0.5 seconds
	BCF	PORTB,0	;Turn OFF B0
	BCF	PORTB,1	;Turn OFF B1
	BCF	PORTB,2	;Turn OFF B2
	BCF	PORTB,3	;Turn OFF B3

```
            BCF        PORTB,4       ;Turn OFF B4
            BCF        PORTB,5       ;Turn OFF B5
            BCF        PORTB,6       ;Turn OFF B6
            BCF        PORTB,7       ;Turn OFF B7
            CALL       DELAYP5       ;Wait 0.5 seconds
            GOTO       BEGIN
END
```

Save the program as FLASH8.ASM, assemble and program the 16F84 as indicated in Chapter 2.

There is an easier way than this of switching all outputs on a port, which we look at later in this chapter with a set of disco lights.

Chasing 8 LEDs

Let's now consider the code to chase the 8 LEDs. The circuit of Figure 3.2 is required for this. The code will switch B0 on and off, then B1, then B2 etc.

```
;Program starts now.

BEGIN       BSF        PORTB,0       ;Turn ON B0
            CALL       DELAYP5       ;Wait 0.5 seconds

            BCF        PORTB,0       ;Turn OFF B0
            BSF        PORTB,1       ;Turn ON B1
            CALL       DELAYP5       ;Wait 0.5 seconds

            BCF        PORTB,1       ;Turn OFF B1
            BSF        PORTB,2       ;Turn ON B2
            CALL       DELAYP5       ;Wait 0.5 seconds

            BCF        PORTB,2       ;Turn OFF B2
            BSF        PORTB,3       ;Turn ON B3
            CALL       DELAYP5       ;Wait 0.5 seconds

            BCF        PORTB,3       ;Turn OFF B3
            BSF        PORTB,4       ;Turn ON B4
            CALL       DELAYP5       ;Wait 0.5 seconds

            BCF        PORTB,4       ;Turn OFF B4
            BSF        PORTB,5       ;Turn ON B5
            CALL       DELAYP5       ;Wait 0.5 seconds
```

```
            BCF      PORTB,5      ;Turn OFF B5
            BSF      PORTB,6      ;Turn ON B6
            CALL     DELAYP5      ;Wait 0.5 seconds


            BCF      PORTB,6      ;Turn OFF B6
            BSF      PORTB,7      ;Turn ON B7
            CALL     DELAYP5      ;Wait 0.5 seconds
            BCF      PORTB,7      ;Turn OFF B7
            CALL     DELAYP5      ;Wait 0.5 seconds
            GOTO     BEGIN
END
```

This code once again is added to the bottom of HEADER84.ASM and is saved as

CHASE8A.ASM

Now that we have chased the LEDs one way let's run them back the other way and call the program CHASE8B.ASM. I think you know the routine add the code to the bottom of HEADER84.ASM etc. So I will not mention it again.

```
;CHASE8B.ASM
;Program starts now.

BEGIN       BSF      PORTB,0      ;Turn ON B0
            CALL     DELAYP5      ;Wait 0.5 seconds


            BCF      PORTB,0      ;Turn OFF B0
            BSF      PORTB,1      ;Turn ON B1
            CALL     DELAYP5      ;Wait 0.5 seconds


            BCF      PORTB,1      ;Turn OFF B1
            BSF      PORTB,2      ;Turn ON B2
            CALL     DELAYP5      ;Wait 0.5 seconds


            BCF      PORTB,2      ;Turn OFF B2
            BSF      PORTB,3      ;Turn ON B3
            CALL     DELAYP5      ;Wait 0.5 seconds


            BCF      PORTB,3      ;Turn OFF B3
            BSF      PORTB,4      ;Turn ON B4
            CALL     DELAYP5      ;Wait 0.5 seconds
```

```
        BCF        PORTB,4        ;Turn OFF B4
        BSF        PORTB,5        ;Turn ON B5
        CALL       DELAYP5        ;Wait 0.5 seconds

        BCF        PORTB,5        ;Turn OFF B5
        BSF        PORTB,6        ;Turn ON B6
        CALL       DELAYP5        ;Wait 0.5 seconds

        BCF        PORTB,6        ;Turn OFF B6
        BSF        PORTB,7        ;Turn ON B7
        CALL       DELAYP5        ;Wait 0.5 seconds

        BCF        PORTB,7        ;Turn OFF B7
        BSF        PORTB,6        ;Turn ON B6
        CALL       DELAYP5        ;Wait 0.5 seconds

        BCF        PORTB,6        ;Turn OFF B6
        BSF        PORTB,5        ;Turn ON B5
        CALL       DELAYP5        ;Wait 0.5 seconds

        BCF        PORTB,5        ;Turn OFF B5
        BSF        PORTB,4        ;Turn ON B4
        CALL       DELAYP5        ;Wait 0.5 seconds

        BCF        PORTB,4        ;Turn OFF B4
        BSF        PORTB,3        ;Turn ON B3
        CALL       DELAYP5        ;Wait 0.5 seconds

        BCF        PORTB,3        ;Turn OFF B3
        BSF        PORTB,2        ;Turn ON B2
        CALL       DELAYP5        ;Wait 0.5 seconds

        BCF        PORTB,2        ;Turn OFF B2
        BSF        PORTB,1        ;Turn ON B1
        CALL       DELAYP5        ;Wait 0.5 seconds
        BCF        PORTB,1        ;Turn OFF B1
        GOTO       BEGIN
END
```

Just one last flasher program. Let us switch each output on in turn leaving them on as we go and then switch them off in turn. Try this for yourselves before looking at the solution!

The program is saved as UPANDDOWN.ASM

;Program starts now.

```
BEGIN     BSF        PORTB,0      ;Turn ON B0
          CALL       DELAYP5      ;Wait 0.5 seconds

          BSF        PORTB,1      ;Turn ON B1
          CALL       DELAYP5      ;Wait 0.5 seconds

          BSF        PORTB,2      ;Turn ON B2
          CALL       DELAYP5      ;Wait 0.5 seconds

          BSF        PORTB,3      ;Turn ON B3
          CALL       DELAYP5      ;Wait 0.5 seconds

          BSF        PORTB,4      ;Turn ON B4
          CALL       DELAYP5      ;Wait 0.5 seconds

          BSF        PORTB,5      ;Turn ON B5
          CALL       DELAYP5      ;Wait 0.5 seconds

          BSF        PORTB,6      ;Turn ON B6
          CALL       DELAYP5      ;Wait 0.5 seconds

          BSF        PORTB,7      ;Turn ON B7
          CALL       DELAYP5      ;Wait 0.5 seconds

          BCF        PORTB,7      ;Turn OFF B6
          CALL       DELAYP5      ;Wait 0.5 seconds

          BCF        PORTB,6      ;Turn OFF B6
          CALL       DELAYP5      ;Wait 0.5 seconds

          BCF        PORTB,5      ;Turn OFF B5
          CALL       DELAYP5      ;Wait 0.5 seconds

          BCF        PORTB,4      ;Turn OFF B4
          CALL       DELAYP5      ;Wait 0.5 seconds

          BCF        PORTB,3      ;Turn OFF B3
          CALL       DELAYP5      ;Wait 0.5 seconds

          BCF        PORTB,2      ;Turn OFF B2
          CALL       DELAYP5      ;Wait 0.5 seconds

          BCF        PORTB,1      ;Turn OFF B1
          CALL       DELAYP5      ;Wait 0.5 seconds
```

BCF	PORTB,0	;Turn OFF B0
CALL	DELAYP5	;Wait 0.5 seconds
GOTO	BEGIN	

END

There are lots of other combinations for you to practice on. I'll leave you to experiment further.

Consider another example of the delay routine:

Traffic lights

If you have ever tried to design a 'simple' set of traffic lights then you will appreciate how much circuitry is required. An oscillator circuit, counters and logic decode circuitry.
The microcontroller circuit is a much better solution even for this 'simple' arrangement. The circuit is shown in Figure 3.3.

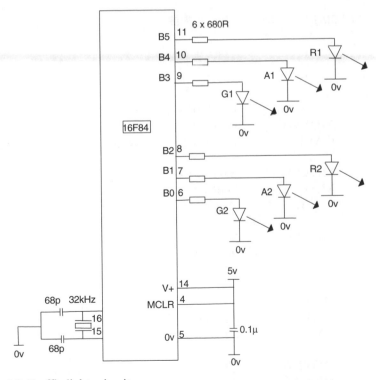

Figure 3.3 Traffic lights circuit

A truth table of the operation of the lights is probably a better aid to a solution rather than a flowchart.

Traffic light truth table

Time	B7	B6	B5	B4	B3	B2	B1	B0
			R1	A1	G1	R2	A2	G2
2sec	0	0	1	0	0	1	0	0
2sec	0	0	1	1	0	1	0	0
5sec	0	0	0	0	1	1	0	0
2sec	0	0	0	1	0	1	0	0
2sec	0	0	1	0	0	1	0	0
2sec	0	0	1	0	0	1	1	0
5sec	0	0	1	0	0	0	0	1
2sec	0	0	1	0	0	0	1	0
REPEAT								

Program listing for the traffic lights

```
;TRAFFIC.ASM
;****************************************************************
;Program starts now.

BEGIN      MOVLW      B'00100100'    ;R1, R2 on.
           MOVWF      PORTB
           CALL       DELAY2         ;Wait 2 Seconds.

           MOVLW      B'00110100'    ;R1, A1, R2 on.
           MOVWF      PORTB
           CALL       DELAY2         ;Wait 2 Seconds.

           MOVLW      B'00001100'    ;G1, R2 on.
           MOVWF      PORTB
           CALL       DELAY5         ;Wait 5 Seconds.

           MOVLW      B'00010100'    ;A1, R2 on.
           MOVWF      PORTB
           CALL       DELAY2         ;Wait 2 Seconds.

           MOVLW      B'00100100'    ;R1, R2 on.
           MOVWF      PORTB
```

```
        CALL        DELAY2          ;Wait 2 Seconds.

        MOVLW       B'00100110'     ;R1, R2, A2 on.
        MOVWF       PORTB
        CALL        DELAY2          ;Wait 2 Seconds.

        MOVLW       B'00100001'     ;R1, G2 on.
        MOVWF       PORTB
        CALL        DELAY5          ;Wait 5 Seconds.

        MOVLW       B'00100010'     ;R1, A2 on.
        MOVWF       PORTB
        CALL        DELAY2          ;Wait 2 Seconds.
        GOTO        BEGIN
END
```

How does it work

In a previous examples we turned LEDs on and off with the two commands BSF and BCF, but a much better way has been used with the TRAFFIC.ASM program.

The basic difference is the introduction of two more commands:

- MOVLW   MOVe the Literal (a number) into the Working register.
- MOVWF   MOVe the Working register to the File.

The data, in this example, binary numbers, are moved to W and then to the file which is the output PORTB to switch the LEDs on and off. Unfortunately the data cannot be placed in PORTB with only one instruction it has to go via the W register.

So:

```
MOVLW   B'00100100'   clears B7,B6, sets B5, clears B4,B3, sets B2
                      and clears B1, B0 in the W register
MOVWF   PORTB         moves the data from the W register to PORTB
                      to turn the relevant LEDs on and off.
```

All 8 outputs are turned on/off with these 2 instructions.

CALL DELAY2 and CALL DELAY5 waits 2 seconds and 5 seconds before continuing with the next operation. DELAY2 and DELAY5 need adding to the subroutine section as:
; 5 second delay.

```
DELAY5 CLRF    TMR0                ;START TMR0.
```

```
LOOPC   MOVF    TMR0,W              ;READ TMR0 INTO W.
        SUBLW   .160                ;TIME - 160
        BTFSS   STATUS,ZEROBIT      ;Check TIME - W = 0
        GOTO    LOOPC               ;Time is not = 160.
        RETLW   0                   ;Time is 160, return.


; 2 second delay.
DELAY2  CLRF    TMR0                ;START TMR0.
LOOPD   MOVF    TMR0,W              ;READ TMR0 INTO W.
        SUBLW   .64                 ;TIME - 64
        BTFSS   STATUS,ZEROBIT      ;Check TIME - W = 0
        GOTO    LOOPD               ;Time is not = 64.
        RETLW   0                   ;Time is 64, return.
```

The W register

The W or working register is the most important register in the micro. It is in the W register were all the calculations and logical manipulations such as addition, subtraction, and-ing, or-ing etc., are done.

The W register shunts data around like a telephone exchange re-routes telephone calls. In order to move data from locationA to locationB, the data has to be moved from locationA to W and then from W to location B

NB. If the three lines in the TRAFFIC.ASM program are repeated then any pattern and any delay can be used to sequence the lights – you can make your own disco lights!

Repetition (e.g. disco lights)

Instead of just repeating one sequence over and over, suppose we wish to repeat several sequences before returning to the start as with a set of disco lights.

Consider the circuit shown in Figure 3.4. The 8 'Disco Lights' B0-B7 are to be run as two sequences.

Sequence 1   Turn all lights on.
             Wait.
             Turn all lights off
             Wait

Sequence 2   Turn B7-B4 ON, B3-B0 OFF
             Wait
             Turn B7-B4 OFF, B3-B0 ON
             Wait

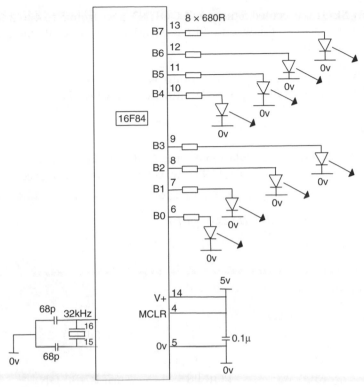

Figure 3.4 Disco lights

Suppose we wish Sequence 1 to run 5 times before going onto Sequence 2 to run 10 times and then repeat. A section of program is repeated a number of times with 4 lines of code shown below:

```
MOVLW    .5          ;Move 5 into W
MOVWF    COUNT       ;Move W into user file COUNT
.
SEQ1
.
DECFSZ   COUNT       ;decrement file COUNT skip if zero.
GOTO     SEQ1        ;COUNT not yet zero, repeat sequence
```

- The first two lines set up a file COUNT with 5. (Count is the first user file and is found in memory location 0CH.) 5 is first of all moved into W then from there to file COUNT.
- SEQ1 is executed.
- The DECFSZ COUNT instruction, DECrement File and Skip if Zero, decrements, takes 1 off, the file COUNT and skips GOTO SEQ1 if the count is zero, if not zero then do SEQ1 again.

This way SEQ1 is executed 5 times and COUNT goes from 5 to 4 to 3 to 2 to 1 to 0 when we skip and follow onto SEQ2. SEQ2 is then done 10 times, say, and the code would be:

```
MOVLW    .10              ;Move 10 into W
MOVWF    COUNT            ;Move W into user file COUNT
.
SEQ2
.
DECFSZ   COUNT            ;decrement file COUNT skip if zero.
GOTO     SEQ2             ;COUNT not yet zero, repeat sequence
```

Program code for the disco lights

```
;DISCO.ASM

;********************************************************
;Program starts now.

BEGIN    MOVLW    .5
         MOVWF    COUNT          ;Set COUNT = 5

SEQ1     MOVLW    B'11111111'
         MOVWF    PORTB          ;Turn B7-B0 ON
         CALL     DELAYP5        ;Wait 0.5 seconds
         MOVLW    B'00000000'
         MOVWF    PORTB          ;Turn B7-B0 OFF
         CALL     DELAYP5        ;Wait 0.5 seconds
         DECFSZ   COUNT          ;COUNT-1, skip if 0.
         GOTO     SEQ1

         MOVLW    .10
         MOVWF    COUNT          ;Set COUNT = 10

SEQ2     MOVLW    B'11110000'
         MOVWF    PORTB          ;B7-B4 on, B3-B0 off
         CALL     DELAYP5        ;Wait 0.5 seconds
         MOVLW    B'00001111'
         MOVWF    PORTB          ;B7-B4 off, B3-B0 on
         CALL     DELAYP5        ;Wait 0.5 seconds

         DECFSZ   COUNT          ;COUNT-1, skip if 0.
         GOTO     SEQ2
         GOTO     BEGIN
END
```

Using the idea of repeating sequences like this any number of combinations can be repeated. The times of course do not need to be of 0.5 seconds duration. The flash rate can be speeded up or slowed down depending on the combination.

Try programming a set of your own Disco Lights. This should keep you quiet for hours (days!).

More than 8 outputs

Suppose we wish to have a set of disco lights in a 3×3 matrix as shown in Figure 3.5. This configuration of course requires 9 outputs. We have 8 outputs on PORTB so we need to make one of the PORTA bits an output also, say PORTA bit0.

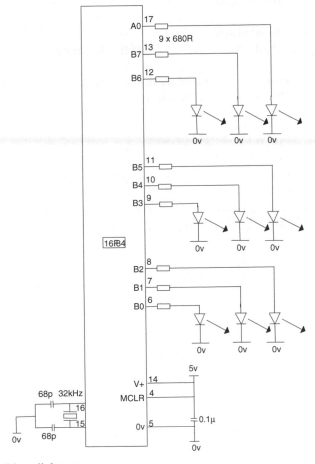

Figure 3.5 9 Disco light set

To change PORTA bit0 from an input to an output change the lines in the Configuration section from:

```
MOVLW      B'00011111'
MOVWF      TRISA
to
MOVLW      B'00011110'
MOVWF      TRISA
```

NB a 1 signifies an input a 0 signifies an output.

So to set a '+' pattern in the lights we turn on B7, B4, B1, B3 and B5, keeping the others off. The code for this would be:

```
MOVLW      B'00000000'
MOVWF      PORTA            ;A0 is clear
MOVLW      B'10111010'
MOVWF      PORTB            ;B7, B5, B4, B3 and B1 are on
```

So to set an 'X' pattern in the lights we turn on B6, B4, B2, A0 and B0, keeping the others off. The code for this would be:

```
MOVLW      B'00000001'
MOVWF      PORTA            ;A0 is on
MOVLW      B'01010101'
MOVWF      PORTB            ;B6, B4, B2 and B0 are on
```

There are endless combinations you can make with 9 lights. In fact there are 512. That is 2^9. This should give you something to go at!

4
Headers, porting code – which micro?

Arizona Microchip the manufacturers of the PIC Microcontroller make over 100 different types of microcontroller. How do we choose the correct one for the job?

Factors affecting the choice of the microcontroller

When deciding on which Microcontroller to use for your application there are a number of factors you will need to consider.

- How many inputs and outputs do you need. If you are using the program FLASHER.ASM which only flashes 1 LED on and off then any PIC will do this. If you are turning 8 outputs on and off then you will need a microcontroller that has at least 8 I/O (of course). So an 8pin micro i.e. 12F629 will not do because it only has 6 I/O.
- Do you need accurate timing? If so then you will need to add a crystal to your micro to provide the clock. If timing is not that critical then you can use a micro that has an on board oscillator such as the 16F818. You can then omit the crystal and 2 capacitors. The timing accuracy is about 1%. This would do for FLASHER.ASM but not for a 24 hour clock. 1% is about 14 minutes a day.
- Are you making analogue measurements? If so you will need a micro with an AtoD converter on it. The 16F818 has a 5 channel, 10 bit AtoD converter. If you need more that 5 channels then you will need to use a micro with more AtoD channels such as the 16F877 which has 8.
- What operating frequency do you require? The greater the frequency the faster your code will execute. Most newer devices can operate up to 20MHz, some even faster. Some older devices can only achieve 4MHz. The programs in this book only require an operating speed of 4MHz.
- How many instructions are there in your program? The 16F818 has space for 1k i.e. 1024 instructions. The 16F877 has 8k program memory locations. All programs in this book require less than 1k of program memory space.
- How many memory locations are required to store data? The 16F818 has 128 bytes of data memory, the 16F877 has 368.

- Do you need to store data so that it will be saved if the power is removed or lost? If so you need a micro with EEPROM data memory. The 16F818 has 128 bytes of EEPROM memory, the 16F877 has 256.

There may be other requirements that you need from your micro, which are not considered in this book, such as:

- Number of timers
- Comparators
- Pulse width modulation
- In circuit debugging
- USB drivers.

Choosing the microcontroller

As I mentioned previously the FLASHER.ASM program which flashes 1 LED on and off can be performed by any Micro. Well, that has narrowed the field down! So which microcontroller do we use for that application? If you were mass producing these flasher units the answer would probably be – use the cheapest and smallest – the 12C508 is possibly the device then. But for small scale production or one offs you will probably have (or develop) a favorite. Probably the most common chip used by the beginner is the 16F84; this has been around since about 1998. This micro has built up a very large fan base which is why it is still widely used. People are using this chip because they are used to using it! There is now another micro on the market which will do everything that the 16F84 can do and more. This device is the 16F818.

The data sheets for the 16F84 and 16F818 are shown in Figures 4.1 and 4.2 respectively.

The main differences are that the 16F818 has 16 I/O, an on board oscillator with 8 selectable frequencies, 128 bytes of data RAM, 128 bytes of EEPROM, 3 Timers one of them a 16 bit, 5 channel 10 bit AtoD converter. The 16F84 has 13 I/O, no on board oscillator, 68 bytes of data RAM, 64 bytes of EEPROM, 1 timer, no AtoD. The most surprising difference of all is that the 16F84 is about 3 times the price of the 16F818!!

The programs in this book consist of 2 parts:

- A header section which tells the 'build' software which device we are using, configures the device, i.e. defines which pins are inputs and outputs, sets the timer rate and includes some timing delays if you require them in a subroutine section.

Devices included in this Data Sheet:

- PIC16F83
- PIC16F84
- PIC16CR83
- PIC16CR84
- Extended voltage range devices available (PIC16**LF**8X, PIC16**LCR**8X)

High Performance RISC CPU Features:

- Only 35 single word instructions to learn
- All instructions single cycle except for program branches which are two-cycle
- Operating speed: DC - 10 MHz clock input
  DC - 400 ns instruction cycle

Device	Program Memory (words)	Data RAM (bytes)	Date EEPROM (bytes)	MAX. Freq (MHz)
PIC16F83	512 Flash	36	64	10
PIC16F84	1 K Flash	68	64	10
PIC16CR83	512 ROM	36	64	10
PIC18CR84	1 K ROM	68	64	10

- 14-bit wide instructions
- 8-bit data path
- 15 special function hardware registers
- Eight-level deep hardware stack
- Direct, indirect and relative addressing modes
- Four interrupt sources:
  - External RB0/INT pin
  - TMR0 timer overflow
  - PORTE<7:4> interrupt on change
  - Data EEPROM write complete
- 1000 erase/write cycles Flash program memory
- 10,000,000 erase/write cycles EEPROM data memory
- EEPROM Data Retention > 40 years

Peripheral Features:

- 13 I/O pins with individual direction control
- High current sink/source for direct LED drive
  - 25 mA sink max. per pin
  - 20 mA source max. per pin
- TMR0: 8-bit timer counter with 8-bit programmable prescaler

Pin Diagrams

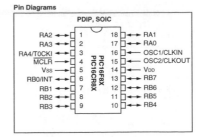

Special Microcontroller Features:

- In-Circuit Serial Programming (ICSP™) - via two pins (ROM devices support only Data EEPROM programming)
- Power-on Reset (POR)
- Power-up Timer (PWRT)
- Oscillator Start-up Timer (OST)
- Watchdog Timer (WDT) with its own on-chip RC oscillator for reliable operation
- Code-protection
- Power saving SLEEP mode
- Selectable oscillator options

CMOS Flash/EEPROM Technology:

- Low-power, high-speed technology
- Fully static design
- Wide operating voltage range:
  - Commercial: 2.0V to 6.0V
  - Industrial: 2.0V to 6.0V
- Low power consumption:
  - < 2 mA typical @ 5V, 4 MHz
  - 15 µA typical @ 2V, 32 kHz
  - < 1 µA typical standby current @ 2V

Figure 4.1 The PIC 16F84 data sheet

- The second part of the program, entitled, 'Program starts now', is where you write the code to perform your application.

The header program is unique to the particular microcontroller being used, but the 'application code' entered after "Program starts now", is specific to the application not the microcontroller. So any microcontroller that has i.e. the required number of I/O or A/D can be used. As I mentioned before any microcontroller can be used to execute the FLASHER.ASM code.

Headers

Just one point before we look at the headers. The 8 pin micros only have 6 I/O, they do not have PORTA and PORTB pins, they have what is called a General

Low-Power Features:
- Power Managed modes:
  - Primary RUN: XT, RC oscillator,
    87 µA, 1 MHz, 2V
  - INTRC: 7 µA, 31.25 kHz, 2V
  - SLEEP: 0.2 µA, 2V
- Timer1 oscillator 1.8 µA, 32 kHz, 2V
- Watchdog Timer: 0.7 µA, 2V
- Wide operating voltage range:
  - Industrial: 2.0V to 5.5V

Oscillators:
- Three Crystal modes:
  – LP, XT, HS: up to 20 MHz
- Two External RC modes
- One External Clock mode:
  - ECIO: up to 20 MHz
- Internal oscillator block:
  - 8 user selectable frequencies: 31 kHz, 125 kHz,
    250 kHz, 500 kHz, 1 MHz, 2 MHz, 4 MHz, 8 MHz

Peripheral Features:
- 16 I/O pins with individual direction control
- High sink/source current: 25 mA
- Timer0: 8-bit timer/counter with 8-bit prescaler
- Timer1: 16-bit timer/counter with prescaler,
  can be incremendet during Sleep via external
  crystal/clock
- Timer2: 8-bit timer/counter with 8-bit period
  register, prescaler and postscaler
- Capature, Compare, PWM (CCP) module:
  - Capature is 16-bit, max. resolution is 12.5 ns
  - Coampare is 16-bit, max. resolution is 200 ns
  - PWM max. resolution is 10-bit
- 10-bit, 5-channel Analog-to-digital converter
- Synchronous Serial Port (SSP) with
  SPI™ (Master/Slave) and I²C™ (Slave)

Pin Diagram

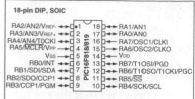

Special Microcontroller Features:
- 100,000 erase/write cycles Enhanced FLASH
  program memory typical
- 1,000,000 typical erase/write cycles EEPROM
  data memory typical
- EEPROM Data Retection: > 40 years
- In-Circuit Serial Proramming™ (ICSP™)-
  via two pins
- Processor read/write access to program memory
- Low Voltage Programming
- In-Circuit Debugging via two pins

Device	Program Memory		Data Memory		I/O Pins	10-bit A/D (ch)	CCP (PWM)	SSP		Timers 8/16-bit
	FLASH (bytes)	# Single Word Instructions	SRAM (bytes)	EEPROM (bytes)				SPI	Slave I²C	
PIC16F818	1792	1024	126	128	16	5	1	Y	Y	2/1
PIC16F819	3584	2048	256	256	16	5	1	Y	Y	2/1

Figure 4.2 The PIC 16F818 and 16F819 data sheet

Purpose I/O or GPIO. So the instruction BSF PORTB,0 would have to be changed to BSF GPIO,0.

The following headers will be used in this book:

HEAD12C508.ASM	; for the 12C508 and 12C509
HEAD12F629.ASM	; for the 12F629
HEAD12F675.ASM	; for the 12F675
HEAD16F627.ASM	; for the 16F627 and 16F628
HEADER84.ASM	; for the 16F84
HEAD16F818.ASM	; for the 16F818 and 16F819
HEAD16F872.ASM	; for the 16F872, 16F874 and 16F877

;**HEAD12C508.ASM** FOR 12C508/9.
;Uses the internal 4MHz clock.

```
TMR0        EQU     1           ;TMR0 is FILE 1.
OSCCAL      EQU     5
GPIO        EQU     6           ;GPIO is FILE 6.
STATUS      EQU     3           ;STATUS is FILE 3.
ZEROBIT     EQU     2           ;ZEROBIT is Bit 2.
COUNT       EQU     07H         ;USER RAM LOCATION.
TIME        EQU     08H         ;TIME IS 39
;*********************************************************
LIST        P = 12C508          ;We are using the 12C508.
ORG         0                   ;0 is the start address.
GOTO        START               ;goto start!
;**************************************************
;Configuration Bits
__CONFIG H'0FEA'                ;selects Internal RC oscillator, WDT off,
                                ;Code Protection disabled.


;*********************************************************
;SUBROUTINE SECTION.

;1/100 SECOND DELAY
DELAY       CLRF    TMR0                        ;START TMR0
LOOPA       MOVF    TMR0,W                      ;READ TMR0 IN W
            SUBWF   TIME,W                      ;TIME - W
            BTFSS   STATUS,ZEROBIT              ;CHECK TIME-W = 0
            GOTO    LOOPA
            RETLW   0                           ;RETURN AFTER TMR0 = 39

;P5 SECOND DELAY
DELAYP5     MOVLW   .50
            MOVWF   COUNT
TIMEC       CALL    DELAY
            DECFSZ  COUNT
            GOTO    TIMEC
            RETLW   0

;1 SECOND DELAY
DELAYP5     MOVLW   .100
            MOVWF   COUNT
TIMED       CALL    DELAY
            DECFSZ  COUNT
            GOTO    TIMED
            RETLW   0
```

```
;**********************************************************
;CONFIGURATION SECTION.

START      MOVWF    OSCCAL
           MOVLW    B'00001000'        ;5 bits of GPIO are O/Ps.
           TRIS     GPIO
           MOVLW    B'00000111'
           OPTION                      ;PRESCALER is /256
           CLRF     GPIO               ;Clears GPIO
           MOVLW    .39
           MOVWF    TIME
;**********************************************************
;Program starts now.
END
```

HEAD12F629.ASM FOR 12F629 using 4MHz internal RC

```
TMR0       EQU      1                  ;TMR0 is FILE 1.
TRISIO     EQU      85H
GPIO       EQU      5                  ;GPIO is FILE 6.
STATUS     EQU      3                  ;STATUS is FILE 3.
ZEROBIT    EQU      2                  ;ZEROBIT is Bit 2.
GO         EQU      1
OPTION_R   EQU      81H
CMCON      EQU      19H
OSCCAL     EQU      90H
COUNT      EQU      20H                ;USER RAM LOCATION.


;**********************************************************
LIST       P = 12F629                  ;We are using the 12F629.
ORG        0                           ;0 is the start address.
GOTO       START                       ;goto start!
;**********************************************************
;Configuration Bits
__CONFIG H'3F84'                       ;selects Internal RC oscillator, WDT off,
                                       ;Code Protection disabled.


;**********************************************************
;SUBROUTINE SECTION.

;1/100 SECOND DELAY
DELAY      CLRF     TMR0               ;START TMR0
LOOPA      MOVF     TMR0,W             ;READ TMR0 IN W
```

```
                SUBLW     .39                    ;TIME - W
                BTFSS     STATUS,ZEROBIT  ;CHECK TIME-W = 0
                GOTO      LOOPA
                RETLW     0                      ;RETURN AFTER TMR0 = 39

;P1 SECOND DELAY
DELAYP1         MOVLW     .10
                MOVWF     COUNT
TIMEC           CALL      DELAY
                DECFSZ    COUNT
                GOTO      TIMEC
                RETLW     0
```

;**
;CONFIGURATION SECTION.

```
START           BSF       STATUS,5        ;BANK1
                MOVLW     B'00001001'     ;BITS 0,3 are I/P
                MOVWF     TRISIO

                MOVLW     B'00000111'
                MOVWF     OPTION_R        ;PRESCALER is /256

                CALL      3FFH
                MOVWF     OSCCAL          ;Calibrates 4MHz oscillator

                BCF       STATUS,5        ;BANK0

                MOVLW     7H
                MOVWF     CMCON           ;Turns off comparator
                CLRF      GPIO            ;Clears GPIO
```

;**
;Program starts now.

END

;**HEAD12F675.ASM** FOR 12F675 using 4MHz internal RC.

```
TMR0       EQU     1        ;TMR0 is FILE 1.
TRISIO     EQU     85H
GPIO       EQU     5        ;GPIO is FILE 6.
STATUS     EQU     3        ;STATUS is FILE 3.
ZEROBIT    EQU     2        ;ZEROBIT is Bit 2.
```

```
GO          EQU         1
ADSEL       EQU         9EH
ADCON0      EQU         1FH
ADRESH      EQU         1EH
OPTION_R    EQU         81H
CMCON       EQU         19H
OSCCAL      EQU         90H
COUNT       EQU         20H             ;USER RAM LOCATION.
```

;***

```
LIST        P = 12F675  ;We are using the 12F675.
ORG         0           ;0 is the start address.
GOTO        START       ;goto start!
```

;***

;Configuration Bits

```
__CONFIG H'3F84'        ;selects Internal RC oscillator, WDT off,
                        ;Code Protection disabled.
```

;***

;SUBROUTINE SECTION.

;1/100 SECOND DELAY
```
DELAY       CLRF        TMR0                    ;START TMR0
LOOPA       MOVF        TMR0,W                  ;READ TMR0 IN W
            SUBLW       .39                     ;TIME - W
            BTFSS       STATUS,ZEROBIT          ;CHECK TIME-W = 0
            GOTO        LOOPA
            RETLW       0                       ;RETURN AFTER TMR0 = 39
```

;P1 SECOND DELAY
```
DELAYP1     MOVLW       .10
            MOVWF       COUNT
TIMEC       CALL        DELAY
            DECFSZ      COUNT
            GOTO        TIMEC
            RETLW       0
```
;***
;CONFIGURATION SECTION.

```
START       BSF         STATUS,5        ;BANK1
            MOVLW       B'00010001'     ;A0 IS ANALOGUE,FOSC/8
            MOVWF       ADSEL
```

```
            MOVLW        B'00001001'      ;BITS 0,3 are I/P
            MOVWF        TRISIO

            MOVLW        B'00000111'
            MOVWF        OPTION_R         ;PRESCALER is /256

            CALL         3FFH
            MOVWF        OSCCAL           ;Calibrates 4MHz oscillator

            BCF          STATUS,5         ;BANK0

            MOVLW        7H
            MOVWF        CMCON            ;Turns off comparator
            CLRF         GPIO             ;Clears GPIO
            BSF          ADCON0,0         ;Turns on A/D converter.
```

;***
;Program starts now.
END

;**HEAD16F627.ASM** for the 16F627/8, using the 37kHz internal RC
;PortA bits 0 to 7 are inputs
;PortB bits 0 to 7 are outputs
;Prescaler/32

;***
;EQUATES SECTION

```
TMR0         EQU          1
OPTION_R     EQU          1
PORTA        EQU          5
PORTB        EQU          6
TRISA        EQU          5
TRISB        EQU          6
STATUS       EQU          3
ZEROBIT      EQU          2
CARRY        EQU          0
EEADR        EQU          1BH
EEDATA       EQU          1AH
EECON1       EQU          1CH
EECON2       EQU          1DH
RD           EQU          0
WR           EQU          1
```

```
WREN          EQU        2
PCON          EQU        0EH
COUNT         EQU        20H
```

;***

```
LIST          P = 16F627    ;using the 627
ORG           0
GOTO          START
```

;***
;Configuration Bits

```
__CONFIG H'3F10'          ;selects Internal RC oscillator, WDT off,
                          ;Code Protection disabled.
```

;***
;SUBROUTINE SECTION.

;0.1 SECOND DELAY
```
DELAYP1   CLRF      TMR0              ;Start TMR0
LOOPA     MOVF      TMR0,W            ;Read TMR0 into W
          SUBLW     .29               ;TIME - W
          BTFSS     STATUS,ZEROBIT    ;Check TIME-W = 0
          GOTO      LOOPA
          RETLW     0                 ;Return after TMR0 = 29
```

;0.5 SECOND DELAY
```
DELAYP5   MOVLW     5
          MOVWF     COUNT
LOOPB     CALL      DELAYP1   ;0.1s delay
          DECFSZ    COUNT
          GOTO      LOOPB
          RETLW     0                 ;Return after 5 DELAYP1
```

;1 SECOND DELAY
```
DELAY1    MOVLW     .10
          MOVWF     COUNT
LOOPC     CALL      DELAYP1   ;0.1s delay
          DECFSZ    COUNT
          GOTO      LOOPC
          RETLW     0                 ;Return after 10 DELAYP1
```

;***

;CONFIGURATION SECTION.

```
START       BSF           STATUS,5        ;Bank1
            MOVLW         B'11111111'
            MOVWF         TRISA           ;PortA is input

            MOVLW         B'00000000'
            MOVWF         TRISB           ;PortB is output

            MOVLW         B'00000100'
            MOVWF         OPTION_R        ;Option Register, TMR0/32
            CLRF          PCON            ;Select 37kHz oscillator.
            BCF           STATUS,5        ;Bank0
            CLRF          PORTA
            CLRF          PORTB
            MOVLW         7
            MOVWF         1FH             ;CMCON turns off comparators.
```

;**
;Program starts now.
END

;**HEADER84.ASM** for the 16F84 using a 32kHz crystal

;EQUATES SECTION

```
TMR0        EQU           1               ;TMR0 is FILE 1.
PORTA       EQU           5               ;PORTA is FILE 5.
PORTB       EQU           6               ;PORTB is FILE 6.
STATUS      EQU           3               ;STATUS is FILE 3.
TRISA       EQU           85H             ;TRISA (the PORTA I/O selection)
TRISB       EQU           86H             ;TRISB (the PORTB I/O selection)
OPTION_R    EQU           81H             ;the OPTION register is file 81H
ZEROBIT     EQU           2               ;ZEROBIT is Bit 2.
COUNT       EQU           0CH             ;USER RAM LOCATION.
```

;**

```
LIST        P = 16F84                     ;We are using the 16F84.
ORG         0                             ;0 is the start address.
GOTO        START                         ;goto start!
```

;**

;Configuration Bits

__CONFIG H'3FF0'   ;selects LP oscillator, WDT off, PUT on,
                                     ;Code Protection disabled.

;***

;SUBROUTINE SECTION.

;1 SECOND DELAY

```
DELAY1     CLRF       TMR0                    ;START TMR0
LOOPA      MOVF       TMR0,W                  ;READ TMR0 IN W
           SUBLW      .32                     ;TIME - W
           BTFSS      STATUS,ZEROBIT  ;CHECK TIME-W = 0
           GOTO       LOOPA
           RETLW      0                       ;RETURN AFTER TMR0 = 32
```

;0.5 SECOND DELAY

```
DELAYP5    CLRF       TMR0                    ;START TMR0
LOOPB      MOVF       TMR0,W                  ;READ TMR0 IN W
           SUBLW      .16                     ;TIME - W
           BTFSS       STATUS,ZEROBIT  ;CHECK TIME-W = 0
           GOTO       LOOPB
           RETLW      0                       ;RETURN AFTER TMR0 = 16
```

;***
;CONFIGURATION SECTION.

```
START      BSF        STATUS,5        ;Turn to BANK1
           MOVLW      B'00011111'     ;5 bits of PORTA are I/Ps.
           MOVWF      TRISA
           MOVLW      B'00000000'
           MOVWF      TRISB           ;PORTB IS OUTPUT

           MOVLW      B'00000111'
           MOVWF      OPTION_R        ;PRESCALER is /256

           BCF        STATUS,5        ;Return to BANK0
           CLRF       PORTA           ;Clears PORTA
           CLRF       PORTB           ;Clears PORTB
           CLRF       COUNT
```
;***
;Program starts now.

END

; **HEAD818.ASM** for 16F818. This sets PORTA as digital INPUT.
;PORTB is an OUTPUT.
;Internal oscillator of 31.25kHz chosen
;The OPTION register is set to /256 giving timing pulses 32.768ms.
;1second and 0.5 second delays are included in the subroutine section.

;***
;EQUATES SECTION

TMR0	EQU	1	;means TMR0 is file 1.
STATUS	EQU	3	;means STATUS is file 3.
PORTA	EQU	5	;means PORTA is file 5.
PORTB	EQU	6	;means PORTB is file 6.
ZEROBIT	EQU	2	;means ZEROBIT is bit 2.
ADCON0	EQU	1FH	;A/D Configuration reg.0
ADCON1	EQU	9FH	;A/D Configuration reg.1
ADRES	EQU	1EH	;A/D Result register.
CARRY	EQU	0	;CARRY IS BIT 0.
TRISA	EQU	85H	;PORTA Configuration Register
TRISB	EQU	86H	;PORTB Configuration Register
OPTION_R	EQU	81H	;Option Register
OSCCON	EQU	8FH	;Oscillator control register.
COUNT	EQU	20H	;COUNT a register to count events.

;***

LIST	P = 16F818	;we are using the 16F818.
ORG	0	;the start address in memory is 0
GOTO	START	;goto start!

;***
;Configuration Bits

__CONFIG H'3F10'   ;sets INTRC-A6 is port I/O, WDT off, PUT
                    ;on, MCLR tied to VDD A5 is I/O
                    ;BOD off, LVP disabled, EE protect disabled,
                    ;Flash Program Write disabled,
                    ;Background Debugger Mode disabled,
                    ;CCP function on B2,
                    ;Code Protection disabled.

;***
;SUBROUTINE SECTION.
;0.1 second delay, actually 0.099968s

DELAYP1	CLRF	TMR0	;START TMR0.
LOOPB	MOVF	TMR0,W	;READ TMR0 INTO W.

```
            SUBLW     .3                    ;TIME-3
            BTFSS     STATUS,ZEROBIT        ;Check TIME-W = 0
            GOTO      LOOPB                 ;Time is not = 3.
            NOP                             ;add extra delay
            NOP
            RETLW     0                     ;Time is 3, return.
```

;0.5 second delay.
```
DELAYP5     MOVLW     .5
            MOVWF     COUNT
LOOPC       CALL      DELAYP1
            DECFSZ    COUNT
            GOTO      LOOPC
            RETLW     0
```

;1 second delay.
```
DELAY1      MOVLW     .10
            MOVWF     COUNT
LOOPA       CALL      DELAYP1
            DECFSZ    COUNT
            GOTO      LOOPA
            RETLW     0
;********************************************************
```

;Configuration Section

```
START       BSF       STATUS,5          ;Turns to Bank1.

            MOVLW     B'11111111'       ;8 bits of PORTA are I/P
            MOVWF     TRISA

            MOVLW     B'00000110'       ;PORTA IS DIGITAL
            MOVWF     ADCON1

            MOVLW     B'00000000'
            MOVWF     TRISB             ;PORTB is OUTPUT

            MOVLW     B'00000000'
            MOVWF     OSCCON            ;oscillator 31.25kHz

            MOVLW     B'00000111'       ;Prescaler is /256
            MOVWF     OPTION_R          ;TIMER is 1/32 secs.
```

```
            BCF            STATUS,5        ;Return to Bank0.
            CLRF           PORTA           ;Clears PortA.
            CLRF           PORTB           ;Clears PortB.
```

;***
;
;Program starts now.
END

;HEAD872.ASM Header for 16F872 using 32kHz oscillator

;EQUATES SECTION

```
            TMR0        EQU        1
            OPTION_R    EQU        1
            PORTA       EQU        5
            PORTB       EQU        6
            PORTC       EQU        7
            TRISA       EQU        5
            TRISB       EQU        6
            TRISC       EQU        7
            STATUS      EQU        3
            ZEROBIT     EQU        2
            CARRY       EQU        0
            EEADR       EQU        0DH
            EEDATA      EQU        0CH
            EECON1      EQU        0CH
            EECON2      EQU        0DH
            RD          EQU        0
            WR          EQU        1
            WREN        EQU        2
            ADCON0      EQU        1FH
            ADCON1      EQU        1FH
            ADRES       EQU        1EH
            CHS0        EQU        3
            GODONE      EQU        2
            COUNT       EQU        20H
```
;***
;
```
            LIST        P = 16F872
            ORG         0
            GOTO        START
```
;***
;
;Configuration Bits
__CONFIG H'3F30'   ;selects LP oscillator, WDT off, PUT on,
                                          ;Code Protection disabled.

```
;************************************************

;SUBROUTINE SECTION.

;1 SECOND DELAY
DELAY1   CLRF      TMR0                    ;Start TMR0
LOOPA    MOVF      TMR0,W                  ;Read TMR0 into W
         SUBLW     .32                     ;TIME - W
         BTFSS     STATUS,ZEROBIT   ;Check TIME-W = 0
         GOTO      LOOPA
         RETLW     0                       ;Return after TMR0 = 32

;0.5 SECOND DELAY
DELAYP5  CLRF      TMR0                    ;Start TMR0
LOOPB    MOVF      TMR0,W                  ;Read TMR0 into W
         SUBLW     .16                     ;TIME - W
         BTFSS     STATUS,ZEROBIT    ;Check TIME-W = 0
         GOTO      LOOPB
         RETLW     0                       ;Return after TMR0 = 16

;**************************************************************
;CONFIGURATION SECTION.

START    BSF           STATUS,5          ;Bank1
         MOVLW         B'11111111'
         MOVWF         TRISA             ;PortA is input

         MOVLW         B'00000000'
         MOVWF         TRISB             ;PortB is output

         MOVLW         B'11111111'
         MOVWF         TRISC             ;PortC is input

         MOVLW         B'00000111'
         MOVWF         OPTION_R          ;Option Register, TMR0/256

         MOVLW         B'00000000'
         MOVWF         ADCON1            ;PortA bits 0,1,2,3,5 are analogue
         BSF           STATUS,6          ;BANK3
         BCF           EECON1,7          ;Data memory on.
         BCF           STATUS,5
         BCF           STATUS,6          ;BANK0 return
         BSF           ADCON0,0          ;turn on A/D
         CLRF          PORTA
```

```
        CLRF        PORTB
        CLRF        PORTC
```

```
;*********************************************************
;Program starts now.
END
```

These headers can be used for applications that use the corresponding microcontrollers. E.g. Any one of them can be used with FLASHER.ASM. Other applications may require functions that are not in all of the devices i.e. AtoD.

The explanation of the operation of the headers will be dealt with later when the individual micros are examined.

5
Using inputs

A control program usually requires more than turning outputs on and off. They switch on and off because an event has happened. This event is then connected to the input of the microcontroller to 'tell' it what to do next. The input could de derived from a switch or it could come from a sensor measuring temperature, light levels, soil moisture, air quality, fluid pressure, engine speed etc.

Analogue inputs are dealt with later, in this chapter we will concern ourselves with digital on/off inputs.

New instructions used in this chapter:
- BTFSC
- BTFSS
- CLRF
- MOVF
- SUBLW
- SUBWF
- RETLW

As an example let us design a circuit so that switch, SW1 will turn an LED on and off.

The circuit diagram is shown in Figure 5.1.

This circuit is using the 16F84 microcontroller with a 32kHz crystal.

It can of course also be performed with any of the microcontrollers discussed previously. Including the 16F818 using its internal oscillator, in which case the crystal and 2×68pF capacitors are not required.

The program to control the hardware would use the following steps:

1. Wait for SW1 to close.
2. Turn on LED1.
3. Wait for SW1 to open.

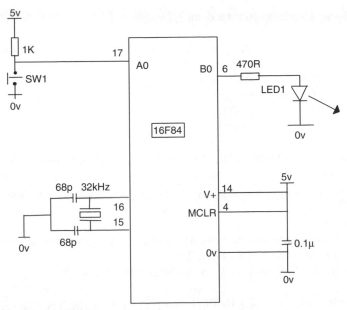

Figure 5.1 Circuit diagram of the microcontroller switch

4. Turn off LED1.
5. Repeat.

In the circuit diagram SW1 is connected to A0 and LED1 to B0.

When the switch is closed A0 goes low or clear. So we wait until A0 is clear. The code for this is:

```
BEGIN    BTFSC      PORTA,0 (test bit 0 in file PORTA skip if clear)
         GOTO       BEGIN
         BSF        PORTB,0
```

- The command BTFSC is Bit Test in File Skip if Clear, and the instruction BTFSC PORTA,0 means Test the Bit in the File PORTA, i.e. Bit0, Skip the next instruction if Clear. If A0 is Clear Skip the next instruction (GOTO BEGIN) if it isn't Clear then do not Skip and GOTO BEGIN to check the switch again.

The program will check the switch thousands maybe millions of times a second, depending on your clock.

- When the switch is pressed the program moves on and executes the instruction BSF PORTB,0 to turn on the LED.

We then wait for the switch to open.

When the switch is open A0 goes Hi or Set, we then wait until A0 is Set i.e.

```
SWOFF    BTFSS      PORTA,0
         GOTO       SWOFF
         BCF        PORTB,0
         GOTO       BEGIN
```

- The command BTFSS is Bit Test in File Skip if Set, and the instruction BTFSS PORTA,0 means Test the Bit in the File PORTA, i.e. Bit0, Skip the next instruction if Set. If A0 is Set Skip the next instruction (GOTO SWOFF) if it isn't Set then do not Skip and GOTO SWOFF to check the switch again.
- When the switch is set the program moves on and executes the instruction BCF PORTB,0 to switch off the LED.
- The program then goes back to the label BEGIN, to repeat.

The program is now added to the header. (NB. Use the TAB to make your listing easy to read.) It is then saved as SWITCH.ASM.

```
;SWITCH.ASM
;**********************************************************
;Program starts now.
BEGIN        BTFSC      PORTA,0      ;Wait for SW1 to be pressed
             GOTO       BEGIN
             BSF        PORTB,0      ;Turn on LED1.
SWOFF        BTFSS      PORTA,0      ;Wait for SW1 to be released.
             GOTO       SWOFF
             BCF        PORTB,0      ;Switch off LED1.
             GOTO       BEGIN        ;Repeat sequence.

END
```

Switch flowchart

It will be obvious from the program listing of the solution to the switch problem that listings are difficult to follow. A picture is worth a thousand words has never been more apt than it is with a program listing. The picture of the program is shown below in the flowchart for the solution to our initial switch problem, Figure 5.2. Before a programming listing is attempted it is very worthwhile drawing a flowchart to depict the program steps. Diamonds are used to show a decision (i.e. a branch) and rectangles are used to show

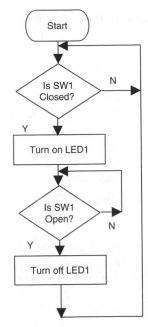

Figure 5.2 Flowchart for the switch

a command. Each shape may take several lines of program to implement. But the idea of the flowchart should be evident. Note that the flowchart describes the problem – you can draw it without any knowledge of the instruction set.

Program development

From our basic switch circuit an obvious addition would be to include a delay so that the LED would go off automatically after a set time.

Suppose we wish to switch the light on for 5 seconds, using A0 as the switch input. Figure 5.3 shows this Delay Flowchart.

The complete listing for this program for the 16F84 is shown below. I have shown the complete code including the header because I have added a 5 second delay in the subroutine section.

```
;DELAY.ASM

;EQUATES SECTION

TMR0      EQU    1      ;TMR0 is FILE 1.
PORTA     EQU    5      ;PORTA is FILE 5.
```

```
PORTB       EQU     6           ;PORTB is FILE 6.
STATUS      EQU     3           ;STATUS is FILE3.
TRISA       EQU     85H         ;TRISA (the PORTA I/O selection)
TRISB       EQU     86H         ;TRISB (the PORTB I/O selection)
OPTION_R    EQU     81H         ;the OPTION register is file 81H
ZEROBIT     EQU     2           ;ZEROBIT is Bit 2.
COUNT       EQU     0CH         ;USER RAM LOCATION.
;********************************************************
LIST        P=16F84             ;We are using the 16F84.
ORG         0                   ;0 is the start address.
GOTO        START               ;goto start!
;********************************************************
;Configuration Bits


__CONFIG H'3FF0'                ;selects LP oscillator, WDT off, PUT on,
                                ;Code Protection disabled.
;********************************************************
;SUBROUTINE SECTION.


;5 second delay.
DELAY5      CLRF    TMR0                    ;Start TMR0.
LOOPA       MOVF    TMR0,W                  ;Read TMR0 into W.
            SUBLW   .160                    ;TIME - 160
            BTFSS   STATUS,ZEROBIT          ;Check TIME-W=0
            GOTO    LOOPA                   ;Time is not=160.
            RETLW   0                       ;Time is 160, return.


;********************************************************
;CONFIGURATION SECTION.


START       BSF     STATUS,5                ;Turn to BANK1
            MOVLW   B'00011111'             ;5 bits of PORTA are I/Ps.
            MOVWF   TRISA
            MOVLW   B'00000000'
            MOVWF   TRISB                   ;PORTB IS OUTPUT
            MOVLW   B'00000111'
            MOVWF   OPTION_R                ;PRESCALER is /256
            BCF     STATUS,5                ;Return to BANK0
            CLRF    PORTA                   ;Clears PORTA
            CLRF    PORTB                   ;Clears PORTB
            CLRF    COUNT


;********************************************************
```

```
;Program starts now.
ON          BTFSC        PORTA,0      ;Check button pressed.
            GOTO         ON
            BSF          PORTB,0      ;Turn on LED.
            CALL         DELAY5       ;CALL 5 second delay
            BCF          PORTB,0      ;Turn off LED.
            GOTO         ON           ;Repeat

END
```

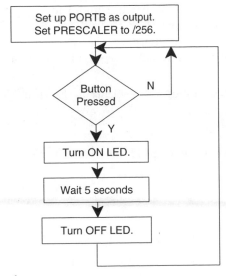

Figure 5.3 Delay flowchart

How does it work?

- We check to see if the switch has been pressed (clear). If not GOTO ON and check again. If it has skip that line and Turn on the LED on B0. The code is:

```
ON          BTFSC        PORTA,0      ;Check button pressed.
            GOTO         ON
            BSF          PORTB,0      ;Turn on LED.
```

- Wait 5 seconds. The 5 second delay has been included for you in the subroutine section. Code:

```
CALL         DELAY5
```

- Turn the LED off and go back to the beginning. Code:

```
BCF          PORTB,0      ;Turn off LED.
GOTO         ON
```

Try this next problem for yourselves, before looking at the solution.

Problem 1:   Using Port A bit 0 as a start button and outputs on PortB
                bits 0-3. Switch on Port B bits 0 and 2 for ¼ second, switch
                off bits 0 and 2.
                Switch on Port B bits 1 and 3 for ¼ second, switch off bits
                1 and 3.
                Repeat continuously.
                The ¼ second delay is provided for you.

The flowchart for the solution to problem1 is shown in Figure 5.4

Program solution to problem1 for the 16F84

```
;PROBLEM1.ASM


;EQUATES SECTION


TMR0         EQU    1        ;TMR0 is FILE 1.
PORTA        EQU    5        ;PORTA is FILE 5.
PORTB        EQU    6        ;PORTB is FILE 6.
STATUS       EQU    3        ;STATUS is FILE 3.
TRISA        EQU    85H      ;TRISA (the PORTA I/O selection)
TRISB        EQU    86H      ;TRISB (the PORTB I/O selection)
OPTION_R     EQU    81H      ;the OPTION register is file 81H
ZEROBIT      EQU    2        ;ZEROBIT is Bit 2.
COUNT        EQU    0CH      ;USER RAM LOCATION.
;*********************************************************
LIST         P = 16F84      ;we are using the 16F84.
ORG          0              ;the start address in memory is 0
GOTO         START          ;goto start!
;*********************************************************
```

;Configuration Bits

__CONFIG H'3FF0'   ;selects LP oscillator, WDT off, PUT on
                                    ;Code Protection disabled.
;***
;SUBROUTINE SECTION.

;0.25 second delay.

```
DELAY     CLRF      TMR0                ;START TMR0.
LOOPA     MOVF      TMR0,W              ;READ TMR0 INTO W.
          SUBLW     .8                  ;TIME - 8
          BTFSS     STATUS,ZEROBIT      ;Check TIME-W = 0
          GOTO      LOOPA               ;Time is not = 8.
          RETLW     0                   ;Time is 8, return.
```

;***

;CONFIGURATION SECTION

```
START     BSF       STATUS,5            ;Turn to BANK1
          MOVLW     B'00011111'         ;5 bits of PORTA are I/Ps.
          MOVWF     TRISA
          MOVLW     B'00000000'
          MOVWF     TRISB               ;PORTB IS OUTPUT
          MOVLW     B'00000111'
          MOVWF     OPTION_R            ;PRESCALER is /256
          BCF       STATUS,5            ;Return to BANK0
          CLRF      PORTA               ;Clears PORTA
          CLRF      PORTB               ;Clears PORTB
```

;***
;Program starts now.

```
ON        BTFSC     PORTA,0       ;Check button pressed.
          GOTO      ON
REPEAT    MOVLW     B'00000101'
          MOVWF     PORTB         ;Turn on bits 0 and 2
          CALL      DELAY         ;¼ second delay
          MOVLW     B'00001010'
          MOVWF     PORTB         ;Turn on bits 1 and 3
          CALL      DELAY         ;¼ second delay
          GOTO      REPEAT        ;Repeat
END
```

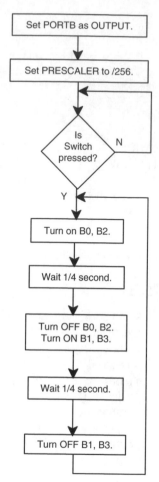

Figure 5.4 Flowchart for problem

How does it work?

- Wait for the switch on PORTA,0 to clear, with BTFSC PORTA,0 then skip to
- MOVLW B'00000101' this sets up the data in the W register.
- MOVWF PORTB transfers the W register to PORTB and puts 5v on B0 and B2 only.
- CALL DELAY waits for ¼ second.
- MOVLW B'00001010' this sets up the data in the W register.
- MOVWF PORTB transfers the W register to PORTB and puts 5v on B1 and B3 only.
- CALL DELAY waits for ¼ second.

- GOTO REPEAT sends the program back to (my) label, REPEAT. This will keep the lights flashing all the time without checking the switch again.

Question. How do we make the program look at the switch, so that we can control whether or not the program repeats?

Answer: Instead of GOTO REPEAT use GOTO BEGIN. The program will then goto the label BEGIN instead of REPEAT and will wait for the switch to be Clear before repeating.

Extra Work. Try and make the flashing routine more interesting by adding more combinations.

Scanning (using multiple inputs)

Scanning (also called polling) is when the microcontroller looks at the condition of a number of inputs in turn and executes a section of program depending on the state of those inputs.

Applications include:

- Burglar Alarms – when sensors are monitored and a siren sounds either immediately or after a delay depending on which input is active.
- Keypad scanning – a key press could cause an LED to light, a buzzer to sound or a missile to be launched. Just do not press the wrong key!

Let's consider a simple example:

Switch scanning

Design a circuit so that if a switch is pressed a corresponding LED will light. i.e.

If SW0 is Hi, (logic1 or Set) then LED0 is on.
If SW0 is Low, (logic 0 or Clear) then LED0 is off.
If SW1 is Hi, (logic1 or Set) then LED1 is on.
If SW1 is Low, (logic 0 or Clear) then LED1 is off.
etc.

The circuit diagram for this is shown if Figure 5.5 and the corresponding flowchart in Figure 5.6.

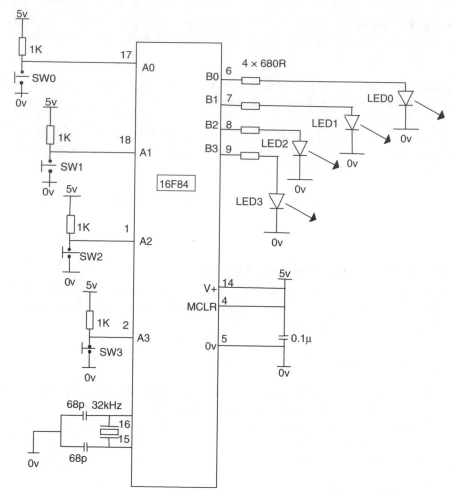

Figure 5.5 Switch scanning circuit

The program for this switch scan is:

;SWSCAN.ASM using 16F84 and 32kHz crystal.

;EQUATES SECTION

```
TMR0        EQU    1      ;TMR0 is FILE 1.
PORTA       EQU    5      ;PORTA is FILE 5.
PORTB       EQU    6      ;PORTB is FILE 6.
```

```
STATUS      EQU    3           ;STATUS is FILE3.
TRISA       EQU    85H         ;TRISA (the PORTA I/O selection)
TRISB       EQU    86H         ;TRISB (the PORTB I/O selection)
OPTION_R    EQU    81H         ;the OPTION register is file 81H
ZEROBIT     EQU    2           ;ZEROBIT is Bit 2.
COUNT       EQU    0CH         ;USER RAM LOCATION.
;****************************************************************
LIST        P=16F84            ;We are using the 16F84.
ORG         0                  ;0 is the start address.
GOTO        START              ;goto start!
;****************************************************************
;Configuration Bits


__CONFIG H'3FF0'               ;selects LP oscillator, WDT off, PUT on,
                               ;Code Protection disabled.


;****************************************************************
;CONFIGURATION SECTION.


START   BSF         STATUS,5           ;Turn to BANK1
        MOVLW       B'00011111'        ;5 bits of PORTA are I/Ps.
        MOVWF       TRISA
        MOVLW       B'00000000'
        MOVWF       TRISB              ;PORTB IS OUTPUT
        MOVLW       B'00000111'
        MOVWF       OPTION_R           ;PRESCALER is /256
        BCF         STATUS,5           ;Return to BANK0
        CLRF        PORTA              ;Clears PORTA
        CLRF        PORTB              ;Clears PORTB
        CLRF        COUNT


;****************************************************************
;Program starts now.

SW0     BTFSC       PORTA,0            ;Switch0 pressed?
        GOTO        TURNON0            ;Yes
        BCF         PORTB,0            :No, Switch off LED0.

SW1     BTFSC       PORTA,1            ;Switch1 pressed?
        GOTO        TURNON1            ;Yes
        BCF         PORTB,1            :NO Switch off LED1.
```

SW2	BTFSC	PORTA,2	;Switch2 pressed?
	GOTO	TURNON2	;Yes
	BCF	PORTB,2	:NO Switch off LED2.
SW3	BTFSC	PORTA,3	;Switch3 pressed?
	GOTO	TURNON3	;Yes
	BCF	PORTB,3	:NO Switch off LED3.
	GOTO	SW0	;Rescan.
TURNON0	BSF	PORTB,0	;Turn on LED0
	GOTO	SW1	
TURNON1	BSF	PORTB,1	;Turn on LED1
	GOTO	SW2	
TURNON2	BSF	PORTB,2	;Turn on LED2
	GOTO	SW3	
TURNON3	BSF	PORTB,3	;Turn on LED3
	GOTO	SW0	

END

How does it work?

- SW0 is checked first with the instruction **BTFSC PORTA,0**. If the switch is closed when the program is executing this line then we **GOTO TURNON0**. That is the program jumps to the label TURNON0 which turns on LED0 and then jumps the program back to check SW1 at, of course, the label, SW1.
- SW1 is then checked in the same manner and then SW2 and SW3.

Suppose we press the switch when the program is not looking at it. The program lines are being executed at ¼ of the clock frequency i.e. 32,768Hz that is 8192 lines a second. The program will always catch you!

Try modifying the program so that the switches can flash 4 different routines e.g. SW0 flashes all lights on and off 5 times for 1 second.

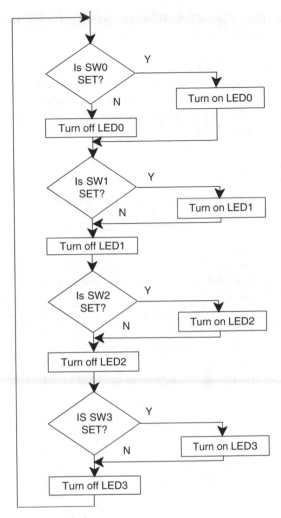

Figure 5.6 Flowchart for switch scan

Control application – a hot air blower

The preceding section outlined how to monitor inputs by looking at them in turn. This application will 'read' all the bits on the port at once, because we will be concerned with particular combinations of inputs rather than individual ones.

The bits on the Input Port will be 0s or 1s and we can treat this binary pattern like any other number in a file.

Consider a controller for a hot air radiator. When the water is warm the fan will blow the warm air into the room. The heater and fan are controlled by 3 temperature sensors: (a) a room temperature sensor, (b) a boiler water temperature sensor and (c) a safety overheating sensor. The truth table for the system is shown in Table 5.1, where a 1 means hot and a 0 means cold for the sensors.

The block diagram for the system is shown in Figure 5.7.

Note A3, A4, A5, A6 and A7 are inputs and need to be connected to 0v. Do not leave them floating – you would not know if they were 0 or 1! Even though

Table 5.1 Truth table for the hot air system

INPUTS								OUTPUTS	
A7	A6	A5	A4	A3	Room A2	Water A1	OverH A0	Heater B1	Fan B0
0	0	0	0	0	0	0	0	1	0
0	0	0	0	0	0	0	1	0	1
0	0	0	0	0	0	1	0	1	1
0	0	0	0	0	0	1	1	0	1
0	0	0	0	0	1	0	0	0	0
0	0	0	0	0	1	0	1	0	1
0	0	0	0	0	1	1	0	0	0
0	0	0	0	0	1	1	1	0	1

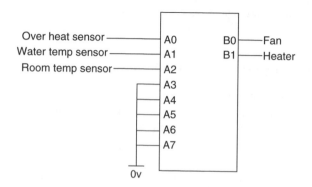

Figure 5.7 Block diagram for the hot air system

they are not being used they are still being read. NB. The inputs A5, A6 and A7 do not exist on the 16F84.

There are 8 input conditions from our 3 sensors. So all 8 must be checked to determine which condition is true.

Consider the first condition $A2 = A1 = A0 = 0$, i.e. PORTA reads 0000 0000. How do we know that PORTA is 0000 0000? We do not have an instruction which says "is PORTA equal to 0000 0000" or any other value for that matter. So we need to look at our 35 instructions and come up with a way of finding out what is the binary value of PORTA.

We check for this condition by subtracting 00000000 from it, if the answer is zero then PORTA reads 00000000. I.e. $0000\ 0000 - 0000\ 0000 = 0$ (obviously). But how do we subtract the two numbers and how do we know if the answer is zero?

This is a very important piece of programming so read the next few lines *carefully*.

- We first of all read PORTA into the W register with the instruction MOVF PORTA,W that moves the data, (setting of the switches, 1s or 0s), into W.
- We then subtract the number we looking for in this case 00000000 from W.
- We then need to know if the answer to this subtraction is zero. If it is then the value on PORTA was 00000000. If the answer is not zero then the value of the data on PORTA was not zero.
- So is the answer zero? Yes or No? The answer is held in a register called the Status Register, in bit 2 of this register, called the zero bit. If the zero bit, called a flag is 1, it is indicating that the statement is true the calculation was zero. If the zero bit is 0 that indicates the statement is false the answer was not zero.
- We test the zero bit in the status register just like we tested the bit on the switch connected to PORTA at the start of this chapter. We use the command BTFSC and the instruction BTFSC STATUS,ZEROBIT. If the zero bit is clear we skip the next instruction if it is set we have a match and do not skip.

The code for this is:

```
MOVLW    B'00000000'        ;put 000000 in W
SUBWF    PORTA              ;subtract W from PORTA
BTFSC    STATUS,ZEROBIT     ;PORTA = 00000000?
CALL     CONDA              ;yes
```

CONDA is short for condition A where we require the heater on and the fan off.

- To check for A2 = A1 = 0 and A0 = 1 we subtract 00000001. To check for the next condition A2 = 0, A1 = 1, A0 = 0 we subtract 00000010, and so on for the other 5 conditions.

```
MOVLW      B'00000001'            ;put 00000001 in W
SUBWF      PORTA                  ;subtract W from PORTA
BTFSS      STATUS,ZEROBIT         ;PORTA = 00000001?
CALL       CONDB                  ;yes
etc.
```

The opcode for this program CONTROL.ASM is:

;CONTROL.ASM

;SUBROUTINE SECTION.

```
CONDA      BCF          PORTB,0        ;turns fan off
           BSF          PORTB,1        ;turns heater on
           RETLW        0

CONDB      BSF          PORTB,0        ;turns fan on
           BCF          PORTB,1        ;turns heater off
           RETLW        0

CONDC      BSF          PORTB,0        ;turns fan on
           BSF          PORTB,1        ;turns heater on
           RETLW        0

CONDD      BCF          PORTB,0        ;turns fan off
           BCF          PORTB,1        ;turns heater off
           RETLW        0
;************************************************************
;Program starts now.

BEGIN      MOVLW        B'00000000'    ;put 00000000 in W
           SUBWF        PORTA          ;PORTA - W
           BTFSC        STATUS,ZEROBIT ;PORTA = 00000000?
           CALL         CONDA          ;yes

           MOVLW        B'00000001'    ;put 00000001 in W
           SUBWF        PORTA          ;PORTA - W
           BTFSC        STATUS,ZEROBIT ;PORTA = 00000001?
           CALL         CONDB          ;yes
```

```
MOVLW    B'00000010'          ;put 00000010 in W
SUBWF    PORTA                ;PORTA - W
BTFSC    STATUS,ZEROBIT       ;PORTA = 00000010?
CALL     CONDC                ;yes

MOVLW    B'00000011'          ;put 00000011 in W
SUBWF    PORTA                ;PORTA - W
BTFSC    STATUS,ZEROBIT       ;PORTA = 00000011?
CALL     CONDB                ;yes

MOVLW    B'00000100'          ;put 00000100 in W
SUBWF    PORTA                ;PORTA - W
BTFSC    STATUS,ZEROBIT       ;PORTA = 00000100?
CALL     CONDD                ;yes

MOVLW    B'00000101'          ;put 00000101 in W
SUBWF    PORTA                ;PORTA - W
BTFSC    STATUS,ZEROBIT       ;PORTA = 00000101?
CALL     CONDB                ;yes

MOVLW    B'00000110'          ;put 00000110 in W
SUBWF    PORTA                ;PORTA - W
BTFSC    STATUS,ZEROBIT       ;PORTA = 00000110?
CALL     CONDD                ;yes

MOVLW    B'00000111'          ;put 00000111 in W
SUBWF    PORTA                ;PORTA - W
BTFSC    STATUS,ZEROBIT       ;PORTA = 00000111?
CALL     CONDB                ;yes

GOTO     BEGIN
```

END

Notice that the SUBROUTINE SECTION needs to be changed to include the conditions, CONDA, CONDB, CONDC and CONDD. The DELAY subroutines are not required in this example.

The program can be checked by using switches for the input sensors and LEDs for the outputs.

There is more than one way of skinning a cat, another way of writing this program is shown in Chapter 8, in the section on look up tables.

6
Understanding the headers

The 16F84

HEADER84.ASM The header for the 16F84.

Now that we have looked at a number of applications we are ready to understand HEADER84.ASM introduced in Chapter 2.

- The header starts with a title that includes the name of the file, this is useful when you are printing it out and details about what the program is doing.

```
;HEADER84.ASM for 16F84.      This sets PORTA as an INPUT (NB 1
;                             means input) and PORTB as an OUTPUT
;                             (NB 0 means output). The OPTION
;                             register is set to /256 to give timing pulses
;                             of 1/32 of a second.
;                             1second and 0.5 second delays are
;                             included in the subroutine section.


;*********************************************************
```

- The EQUATES section tells the software what numbers your words represent. When you write your program you use mnemonics such as PORTA, PORTB, TMR0, STATUS, ZEROBIT, COUNT, MYAGE. The Assembler Program does not understand your words; it is looking for the file number or the bit number. You have to tell it what these mean in the Equates Section i.e. COUNT is File 0C, PortA is file 5, the STATUS register is file 3, ZEROBIT is bit 2, etc. The memory map of the 16F84 in Table 6.1 shows the addresses of the registers and user files. The file with address 0C is the first of the user files and I have called it COUNT, it stores the number of times certain events have happened in my program.

I could have file 0D as COUNT2, file 0E as COUNT3, file 0F as SECONDS or WAIT etc.

;EQUATES SECTION

TMR0	EQU	1	;TMR0 is FILE 1.
PORTA	EQU	5	;PORTA is FILE 5.
PORTB	EQU	6	;PORTB is FILE 6.
STATUS	EQU	3	;STATUS is FILE 3.
TRISA	EQU	85H	;TRISA (the PORTA I/O selection)
TRISB	EQU	86H	;TRISB (the PORTB I/O selection)
OPTION_R	EQU	81H	;the OPTION register is file 81H
ZEROBIT	EQU	2	;ZEROBIT is Bit 2.
COUNT	EQU	0CH	;USER RAM LOCATION.

- What chip are we using?

LIST	P=16F84	;we are using the 16F84.
ORG	0	;the start address in memory is 0
GOTO	START	;goto start!

LIST P=16F84 tells the assembler what chip to assemble the code for. ORG 0 means put the next line of code into program memory address 0, then follow with next line in address1 etc.

GOTO START makes the program bypass the subroutine section and GOTO the label START which is where the device is configured before executing the body of the program. The instruction GOTO START is placed in EPROM address 0 by ORG 0.

The line DELAY1 CLRF TMR0 is then placed in program memory address 1, etc.

- CONFIGURATION BITS
To avoid having to set the configuration bits when we come to program the device they can be set in the code. You can change these bits if you require in MPLAB, note the new number and substitute it in the code.
; Configuration Bits

__CONFIG H'3FF0'   ;selects LP oscillator, WDT off, PUT on,
                      ;Code Protection disabled.

- SUBROUTINE SECTION.
The subroutine section consists of 2 subroutines DELAY1 and DELAYP5.

A subroutine is a section of program, which is, used a number of times instead of rewriting it and using up program memory. Just call it i.e. CALL

DELAY1, at the end you RETURN to the program in the position you left it. The stack is the register that remembers where you came from and returns you back.

The DELAY1 code is:

```
DELAY1    CLRF     TMR0                    ;Start TMR0.
LOOPA     MOVF     TMR0,W                  ;Read TMR0 into W.
          SUBLW    .32                     ;TIME - 32
          BTFSS    STATUS,ZEROBIT          ;Check TIME-W = 0
          GOTO     LOOPA                   ;Time is not = 32.
          RETLW    0                       ;Time is 32, return.
```

DELAY1 starts by clearing the register TMR0 (timer 0), with CLRF TMR0, i.e. CleaR the File TMR0. This sets the timer to zero and will be counting TMR0 pulses every 1/32 of a second.

LOOPA MOVF TMR0,W is move file TMR0 into the working register, W.

We want to know when TMR0 is 32, because then we will have had 32 TIMER0 pulses, which is 1 second. This is done with a subtraction as in the example earlier in this Chapter 5, in the section on the hot air blower.

The label LOOPA is there because we keep returning to it until TMR0 reaches the required value.

There is no instruction, which asks the micro is TMR0 equal to 32. So we have to use the instructions available. We subtract a number from W and ask is the answer 0. If for example we subtract 135 from W and the answer is 0 then W contained 135 if the answer was not 0 then W did not contain 135. The status register contains a bit called a zerobit, it is bit2. Notice in the EQUATES section I have put ZEROBIT EQU 2. So I can use ZEROBIT in my code instead of 2 — I would soon forget what the 2 was supposed to mean. The zerobit is set to a 1 when the result of a previous calculation is 0. So a 1 means result was 0!!!! Think of this as a flag (because that's what it is called), the flag is waving (a 1) to indicated the result is zero. We can test this zerobit, i.e. look at it and see if it is a 1 or 0. We can skip the next instruction if it is set, (a zero has occurred), by BTFSS STATUS,ZEROBIT or skip if clear, (a zero has not occurred), by BTFSC STATUS, ZEROBIT. Doesn't this read better than BTFSC 3,2 STATUS is Register3, ZEROBIT is bit 2.

Lets look at this subroutine again.

```
DELAY1    CLRF      TMR0                  ;START TMR0.
LOOPA     MOVF      TMR0,W                ;READ TMR0 INTO W.
          SUBLW     .32                   ;TIME - 32
          BTFSS     STATUS,ZEROBIT        ; Check TIME-W = 0
          GOTO      LOOPA                 ;Time is not = 32.
          RETLW     0                     ;Time is 32, return.
```

- We clear TMR0 (CLRF TMR0).
- Then move TMR0 into W (MOVF TMR0,W)
- SUBTRACT 32 from W which now holds TMR0 value. (SUBLW .32)
- If W (hence TMR0) is 32 the zerobit is set, we skip the next instruction and return from the subroutine with 0 in W (RETLW 0)
- If W is not 32 then we do not skip and we GOTO LOOPA and put TMR0 in W and repeat until TMR0 is 32.

DELAYP5 is a similar code but TMR0 now is only allowed to count upto 16 i.e. a half-second (with 32 pulses a second). Note if you copy and paste, change the name of the subroutine from DELAY1 to DELAYP5, change the 32 to 16 and do not forget to change LOOPA to LOOPB. You cannot goto room 27 if there are two room 27s!

- CONFIGURATION SECTION:

```
START     BSF       STATUS,5              ;Turn to BANK1
          MOVLW     B'00011111'           ;5 bits of PORTA are I/Ps.
          MOVWF     TRISA
          MOVLW     B'00000000'
          MOVWF     TRISB                 ;PORTB IS OUTPUT
          MOVLW     B'00000111'
          MOVWF     OPTION_R              ;PRESCALER is /256
          BCF       STATUS,5              ;Return to BANK0
          CLRF      PORTA                 ;Clears PORTA
          CLRF      PORTB                 ;Clears PORTB
          CLRF      COUNT
```

The instruction BSF STATUS,5 sets bit 5 in the Status Register. As you can see from the explanation of the Status Register bits in Chapter 19, bit 5 is a page select bit which selects page1 giving us access to the Registers in the page 1 (Bank1) column of the memory map in Table 6.1. The reason for pages or banks is that we have an 8 bit micro. 8 bits can only address 256 files so to identify a file we have it on a page, like a line in a book i.e. line 17 on page 40 instead of line 2475.

```
MOVLW      B'00011111'              ;5bits of PORTA are I/P
MOVWF      TRISA
```

These 2 lines move 11111 into the data direction register to set the 5 bits of PORTA as inputs. The 11111 is first moved to W (MOVLW B'00011111') and then into the data direction register with MOVWF TRISA. A 1 signifies an input a 0 an output.

```
MOVLW      B'00000000'              ;8bits of PORTB are O/P
MOVWF      TRISB
```

These 2 lines move 00000000 into the data direction register to set the 8 bits of PORTB as outputs. The 000000 is first moved to W and then into the data direction register with MOVWF TRISB.

PortA and PortB can be configured differently if required. E.g. to make the lower 4 bits of PortB outputs and the upper 4 bits inputs - alter the 2 lines of the program with:

```
           MOVLW      B'11110000'
           MOVWF      TRISB
```

The header also sets the internal clock to divide by 256 i.e. a 32.768kHz clock gives a program execution of 32.768 kHz/4 = 8.192 kHz. If the prescaler is set to divide by 256 this gives timing pulses of 32 a second.

The prescaler is configured with the 2 lines:

```
MOVLW      B'00000111'              ;Prescaler is /256
MOVWF      OPTION_R                 ;TIMER is 1/32 sec.
```

The OPTION register can be altered in the header to give faster timing pulses if required, as described in the OPTION Register section in Chapter 19.

```
The line     BCF STATUS,5           ;Return to Bank0.
```

then returns to page 0 on the memory map. The good news here is in the programs in this book we only need to go into page 1 in the Configuration Section. The body of the program, your section, resides in page 0.

We then finish the configuration section by clearing any outputs in PORTA and PORTB with,

CLRF   PORTA   ;Clears PortA.
CLRF   PORTB   ;Clears PortB.

This will not affect any bits that are configured as inputs.

Just for good measure the COUNT file is also cleared with CLRF COUNT.

16F84 memory map

The Memory Map of the 16F84 is shown in Table 6.1.

This diagram shows the position of the Special Function Registers, i.e. PORTA, PORTB, TMR0 etc. in addresses 00 to 0B and the location of the User Files i.e. COUNT (the only one we have used up to now) occupying locations 0C through to 4F.

These files are very important when writing our code. The Special Function Registers enable us to tell the microcontroller to do things, i.e. set PORTB up as an output port with TRISB, alter the rate of TMR0 with the OPTION

Table 6.1 16F84 memory map

FILE ADDRESS	FILE NAME	FILE NAME
00	INDIRECT ADDRESS	INDIRECT ADDRESS
01	TMR0	OPTION
02	PCL	PCL
03	STATUS	STATUS
04	FST	FSR
05	PORTA	TRISA
06	PORTB	TRISB
07	-	-
08	EEDATA	EECON1
09	EDADR	EECON2
0A	PCLATH	PCLATH
0B	INTCON	INTCON
0C	68	
	USER	
4F	FILES	
	BANK0	BANK1

register, find out if the result of a calculation is zero, +ve or −ve using the STATUS register. TMR0 of course tells us how much time has elapsed.

The other microcontroller which features frequently in this book, my favourite, is the 16F818. We will look at its header and memory map now and compare it to the 16F84 to see how they differ. After that you will be able to distinguish between other micros.

The 16F818

HEAD818.ASM The header for the 16F818.

The code shown below is the header for the 16F818 that we first saw in Chapter 4.

```
;HEAD818.ASM for 16F818. This sets PORTA as digital INPUT.
;PORTB is an OUTPUT.
;Internal oscillator of 31.25kHz chosen
;The OPTION register is set to /256 giving timing pulses of 32.768 ms.
;1second and 0.5 second delays are included in the subroutine section.

;********************************************************

; EQUATES SECTION

TMR0       EQU   1       ;means TMR0 is file 1.
STATUS     EQU   3       ;means STATUS is file 3.
PORTA      EQU   5       ;means PORTA is file 5.
PORTB      EQU   6       ;means PORTB is file 6.
ZEROBIT    EQU   2       ;means ZEROBIT is bit 2.
ADCON0     EQU   1FH     ;A/D Configuration reg.0
ADCON1     EQU   9FH     ;A/D Configuration reg.1
ADRES      EQU   1EH     ;A/D Result register.
CARRY      EQU   0       ;CARRY IS BIT 0.
TRISA      EQU   85H     ;PORTA Configuration Register
TRISB      EQU   86H     ;PORTB Configuration Register OPTION_R
OPTION_R   EQU   81H     ;Option Register
OSCCON     EQU   8FH     ;Oscillator control register.
COUNT      EQU   20H     ;COUNT a register to count events.
;********************************************************

LIST       P=16F818      ;we are using the 16F818.
ORG        0             ;the start address in memory is 0
GOTO       START         ;goto start!

;********************************************************
```

;Configuration Bits

```
__CONFIG H'3F10'          ;sets INTRC-A6 is port I/O, WDT off, PUT
                          ;on, MCLR tied to VDD A5 is I/O
                          ;BOD off, LVP disabled, EE protect disabled,
                          ;Flash Program Write disabled,
                          ;Background Debugger Mode disabled, CCP
                          ;function on B2,
                          ;Code Protection disabled.
```

;**
;SUBROUTINE SECTION.

```
;0.1 second delay, actually 0.099968s
DELAYP1   CLRF     TMR0                    ;START TMR0.
LOOPB     MOVF     TMR0,W                  ;READ TMR0 INTO W.
          SUBLW    .3                      ;TIME - 3
          BTFSS    STATUS,ZEROBIT          ; Check TIME-W = 0
          GOTO     LOOPB                   ;Time is not = 3.
          NOP                              ;add extra delay
          NOP
          RETLW 0                          ;Time is 3, return.

;0.5 second delay.
DELAYP5   MOVLW    .5
          MOVWF    COUNT
LOOPC     CALL     DELAYP1
          DECFSZ   COUNT
          GOTO     LOOPC
          RETLW    0

;1 second delay.
DELAY1    MOVLW    .10
          MOVWF    COUNT
LOOPA     CALL     DELAYP1
          DECFSZ   COUNT
          GOTO     LOOPA
          RETLW    0
```

;***
;CONFIGURATION SECTION.

```
START     BSF      STATUS,5               ;Turns to Bank1.

          MOVLW    B11111111'             ;8 bits of PORTA are I/P
          MOVWF    TRISA
```

```
        MOVLW       B'00000110'     ;PORTA IS DIGITAL
        MOVWF       ADCON1


        MOVLW       B'00000000'
        MOVWF       TRISB           ;PORTB is OUTPUT


        MOVLW       B'00000000'
        MOVWF       OSCCON          ;oscillator 31.25kHz


        MOVLW       B'00000111'     ;Prescaler is /256
        MOVWF       OPTION_R        ;TIMER is 1/32 secs.


        BCF         STATUS,5        ;Return to Bank0.
        CLRF        PORTA           ;Clears PortA.
        CLRF        PORTB           ;Clears PortB.


;********************************************************
;Program starts now.
END
```

We will now consider only the new additions to the previous HEADER84. ASM for the 16F84.

- PORTA is now an 8 I/O port, NB. PORTA,5 is input only.
- ADCON0, ADCON1 and ADRES are Special Function Registers that will enable us to instruct the microcontroller on how we want the A/D converter to function. We will discuss these when we consider A/D conversion in Chapter 11.
- OSCCON allows us to set the value of the internal oscillator. We can choose from 8MHz, 4MHz, 2MHz, 1MHz, 500kHz, 250kHz, 125kHz or 31.25 kHz. The use of OSCCON is described in the Register section in Chapter 19.
- CONFIGURATION BITS. There are more functions on the 16F818 than the 16F84 so there are more choices in the way it is configured. Here we have selected the internal oscillator so we do not need the crystal, that has freed up 2 I/O lines. The master clear, MCLR has been switched internally to Vdd (5v) freeing up another I/O line, giving 16 I/O. We have switched the brown out off this would reset the micro if the supply voltage fell below a critical point avoiding erratic behaviour. Low voltage programming has been switched off. EEPROM protection and Program Write Protection has been disabled. Background Debugger Mode has been disabled. The 16F818 is capable of working with the Microchip In Circuit Debugger (ICD2). Capture and Compare

Pulse Width Module (CCP) not discussed in this book has been switched onto B2.

- SUBROUTINE SECTION.
The 16F818 header described uses the internal 31.25kHz oscillator, which does not lend itself so easily to times of seconds. I have had to write a different code for the delays. A 31.25kHz clock gives timing pulses of 32µs which do not add up exactly to give a second. The delay loop similar in its action to the 16F84 delay has had 2 NOP (no operation) instructions added to make up the shortfall. The 0.1 second delay is therefore 0.099968s which is as close as I could get it. If you really need accurate times you will need to use a crystal for your timing. The internal oscillators are only about 1% accurate.

- CONFIGURATION SECTION.
Because the 16F818 has an A/D converter on board you need to tell it which PORTA inputs are analogue and which are digital. Analogue inputs are dealt with in Chapter 11 for now PORTA has been set to all digital inputs with:

| MOVLW | B'00000110' | ;PORTA IS DIGITAL |
| MOVWF | ADCON1 | |

The internal oscillator is set to 31.25kHz with:

| MOVLW | B'00000000' | |
| MOVWF | OSCCON | ;oscillator 31.25kHz |

This is a default condition and is therefore not required. I have included it incase you are wondering how the frequency is set. You need to alter the data in OSCCON to change the frequency, see Chapter 19.

Because the 16F818 has more functions than the 16F84 it follows that there are more Special Function Registers to handle these extra functions. It also has more user files.

These files are now arranged over 4 banks, BANK0, BANK1, BANK2 and BANK3. The Banks are selected by the Bank Select bits (page select bits) in the Status Register, RP0 and RP1, bits 5 and 6, shown in Figure 6.1.

IRP	RP1	RP0	TO	PD	Z	DC	C
bit7							bit0

Figure 6.1 Status register bits

So 00 selects Bank0
 01 selects Bank1
 10 selects Bank2
 11 selects Bank3

For most applications in this book once we have configured the device we will not need to change banks. The only time we do change is when we look at applications involving the Data EEPROM.

The 16F818 memory Map is shown below in Figure 6.2.

	File Address		File Address		File Address		File Address
Indirect addr.(*)	00h	Indirect addr.(*)	80h	Indirect addr.(*)	100h	Indirect addr.(*)	180h
TMR0	01h	OPTION	81h	TMRC	101h	OPTION	181h
PCL	02h	PCL	82h	PCL	102h	PCL	182h
STATUS	03h	STATUS	83h	STATUS	103h	STATUS	183h
FSR	04h	FSR	84h	FSR	104h	FSR	184h
PORTA	05h	TRISA	85h		105h		185h
PORTB	06h	TRISB	86h	PORTB	106h	TRISB	186h
	07h		87h		107h		187h
	08h		88h		108h		188h
	09h		89h		109h		189h
PCLATH	0Ah	PCLATH	8Ah	PCLATH	10Ah	PCLATH	18Ah
INTCON	0Bh	INTCON	8Bh	INTCON	10Bh	INTCON	18Bh
PIR1	0Ch	PIE1	8Ch	EEDATA	10Ch	EECON1	18Ch
PIR2	0Dh	PIE2	8Dh	EEADR	10Dh	EECON2	18Dh
TMR1L	0Eh	PCON	8Eh	EEDATH	10Eh	Reserved[1]	18Eh
TMR1H	0Fh	OSCCON	8Fh	EEADRH	10Fh	Reserved[1]	18Fh
T1CON	10h	OSCTUNE	90h		110h		190h
TMR2	11h		91h				
T2CON	12h	PR2	92h				
SSPBUF	13h	SSPADD	93h				
SSPCON	14h	SSPSTAT	94h				
CCPR1L	15h		95h				
CCPR1H	16h		96h				
CCP1CON	17h		97h				
	18h		98h				
	19h		99h				
	1Ah		9Ah				
	1Bh		9Bh				
	1Ch		9Ch				
	1Dh		9Dh				
ADRESH	1Eh	ADRESL	9Eh				
ADCON0	1Fh	ADCON1	9Fh		11Fh		19Fh
	20h	General Purpose Register	A0h		120H		1A0h
General Purpose Register				accesses 20h-7Fh		accesses 20h-7Fh	
		32 Bytes	BFh				
			C0h				
96 Bytes		accesses 40h-7Fh					
	7Fh		FFh		17Fh		1FFh
Bank 0		Bank 1		Bank 2		Bank 3	

Figure 6.2 16F818 memory map

We will now continue with some more applications and introduce some more instructions and ideas. Each of these programs will be able to be executed using a number of micros using the appropriate headers.

7
Keypad scanning

There are no new instructions used in this chapter

Keypads are an excellent way of entering data into the microcontroller. The keys are usually numbered but they could be labeled as function keys for example in a remote control handset in a TV to adjust the sound or colour etc.

As well as remote controls, keypads find applications in burglar alarms, door entry systems, calculators, microwave ovens etc. So there are no shortage of applications for this section.

Keypads are usually arranged in a matrix format to reduce the number of I/O connections.

A 12 key keypad is arranged in a 3×4 format requiring 7 connections.
A 16 key keypad is arranged in a 4×4 format requiring 8 connections.

Consider the 12 key keypad. This is arranged in 3 columns and 4 rows as shown in Table 7.1. There are 7 connections to the keypad – C1, C2, C3, R1, R2, R3 and R4.

Table 7.1 12 Key keypad

	Column1, C1	Column2, C2	Column3, C3
Row1, R1	1	2	3
Row2, R2	4	5	6
Row3, R3	7	8	9
Row4, R4	*	0	#

This connection to the micro is shown in Figure 7.1.

The keypad works in the following way:

If for example key 6 is pressed then B2 will be joined to B4. For key 1 B0 would be joined to B3 etc. as shown in Figure 7.1.

The micro would set B0 low and scan B3, B4, B5 and B6 for a low to see if keys 1, 4, 7 or * had been pressed.

The micro would then set B1 low and scan B3, B4, B5 and B6 for a low to see if keys 2, 5, 8 or 0 had been pressed.

Finally B2 would be set low and B3, B4, B5 and B6 scanned for a low to see if keys 3, 6, 9 or # had been pressed.

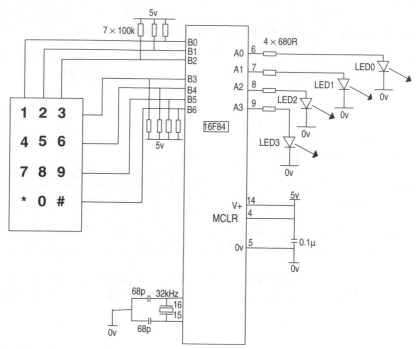

Figure 7.1 Keypad connection to the microcontroller

Programming example for the keypad

As a programming example when key 1 is pressed display a binary 1 on PORTA, when key 2 is pressed display a binary 2 on PORTA etc.

Key 0 displays 10. Key * displays 11. Key # displays 12.

This program could be used as a training aid for decimal to binary conversion.

The flowchart is shown in Figure 7.2.

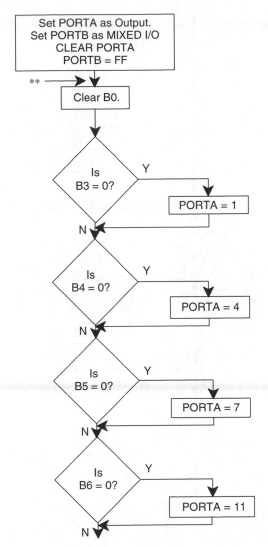

Figure 7.2 Keypad scanning flowchart

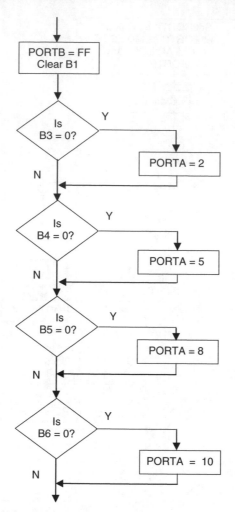

Figure 7.2 Continued

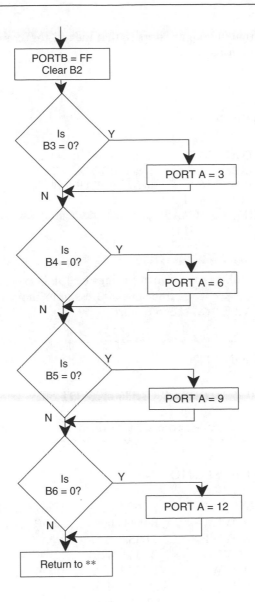

The program listing for the Keypad example for the 16F84 is shown below but can be used with any 'suitable' microcontroller using the appropriate header.

N.B. PORTA has been configured as an output port and PORTB has been configured with 3 outputs and 5 inputs, so the header will require modifying as shown.

PORTB has internal pull up resistors so that the resistors connected to PORTB in Figure 7.1 are not required.

```
;KEYPAD.ASM

;EQUATES SECTION

STATUS       EQU    3       ;means STATUS is file 3.
PORTA        EQU    5       ;means PORTA is file 5.
PORTB        EQU    6       ;means PORTB is file 6.
TRISA        EQU    85H
TRISB        EQU    86H
OPTION_R     EQU    81H


;********************************************************
LIST         P = 16F84      ;we are using the 16F84.
ORG          0              ;the start address in memory is 0
GOTO         START          ;goto start!

;********************************************************
;CONFIGURATION BITS

__Config H'3FF0'    ;selects LP Oscillator, WDT off,
                    ;Put on,
                    ;code protection disabled.
;********************************************************

;CONFIGURATION SECTION

START        BSF         STATUS,5       ;Turns to Bank1.
             MOVLW       B'00000000'    ;PORTA is OUTPUT
             MOVWF       TRISA
             MOVLW       B'11111000'
             MOVWF       TRISB          ;PORTB is mixed I/O.
             BCF         OPTION_R,7     ;Turn on pull ups.
             BCF         STATUS,5       ;Return to Bank0.
             CLRF        PORTA          ;Clears PortA.
             CLRF        PORTB          ;Clears PortB.
;********************************************************
;Program starts now.

COLUMN1      BCF         PORTB,0        ;Clear B0
             BSF         PORTB,1        ;Set B1
             BSF         PORTB,2        ;Set B2
```

```
CHECK1      BTFSC      PORTB,3      ;Is B3 Clear?
            GOTO       CHECK4       ;No
            MOVLW      .1           ;Yes, output 1.
            MOVWF      PORTA
CHECK4      BTFSC      PORTB,4      ;Is B4 Clear?
            GOTO       CHECK7       ;No
            MOVLW      .4           ;Yes, output 4.
            MOVWF      PORTA
CHECK7      BTFSC      PORTB,5      ;Is B5 Clear?
            GOTO       CHECK11      ;No
            MOVLW      .7           ;Yes, output 7.
            MOVWF      PORTA
CHECK11     BTFSC      PORTB,6      ;Is B6 Clear?
            GOTO       COLUMN2      ;No
            MOVLW      .11          ;Yes, output 11.
            MOVWF      PORTA

COLUMN2     BSF        PORTB,0      ;Set B0
            BCF        PORTB,1      ;Clear B1
            BSF        PORTB,2      ;Set B2
CHECK2      BTFSC      PORTB,3      ;Is B3 Clear?
            GOTO       CHECK5       ;No
            MOVLW      .2           ;Yes, output 2.
            MOVWF      PORTA
CHECK5      BTFSC      PORTB,4      ;Is B4 Clear?
            GOTO       CHECK8       ;No
            MOVLW      .5           ;Yes, output 5.
            MOVWF      PORTA
CHECK8      BTFSC      PORTB,5      ;Is B5 Clear?
            GOTO       CHECK10      ;No
            MOVLW      .8           ;Yes, output 8.
            MOVWF      PORTA
CHECK10     BTFSC      PORTB,6      ;Is B6 Clear?
            GOTO       COLUMN3      ;No
            MOVLW      .10          ;Yes, output 10.
            MOVWF      PORTA

COLUMN3     BSF        PORTB,0      ;Set B0
            BSF        PORTB,1      ;Set B1
            BCF        PORTB,2      ;Clear B2
CHECK3      BTFSC      PORTB,3      ;Is B3 Clear?
            GOTO       CHECK6       ;No
            MOVLW      .3           ;Yes, output 3.
            MOVWF      PORTA
CHECK6      BTFSC      PORTB,4      ;Is B4 Clear?
            GOTO       CHECK9       ;No
            MOVLW      .6           ;Yes, output 6.
            MOVWF      PORTA
```

```
CHECK9      BTFSC     PORTB,5      ;Is B5 Clear?
            GOTO      CHECK12      ;No
            MOVLW     .9           ;Yes, output 9.
            MOVWF     PORTA
CHECK12     BTFSC     PORTB,6      ;Is B6 Clear?
            GOTO      COLUMN1      ;No
            MOVLW     .12          ;Yes, output 12.
            MOVWF     PORTA
            GOTO      COLUMN1      ;Start scanning again.

END
```

How does the program work?

Port configuration

The first thing to note about the keypad circuit is that the PORTA pins are being used as outputs. On PORTB, pins B0, B1 and B2 are outputs and B3, B4, B5 and B6 are inputs. So PORTB is a mixture of inputs and outputs. The HEADER84.ASM program has to be modified to change to this new configuration.

To change PORTA to an output port, the following two lines are used in the Configuration Section:

```
MOVLW     B'00000000'  ;PORTA is OUTPUT
MOVWF     TRISA
```

To configure PORTB as a mixed input and output port the following two lines are used in the Configuration Section:

```
MOVLW     B'11111000'
MOVWF     TRISB        ;PORTB is mixed I/O. B0,B1,B2 are O/P.
```

Scanning routine

The scanning routine looks at each individual key in turn to see if one is being pressed. Because it can do this so quickly it will notice we have pressed a key even if we press it quickly.

The scanning routine first of all looks at the keys in column1 i.e. 1, 4, 7 and *. It does this by setting B0 low, B1 and B2 high. If a 1 is pressed the B3 will be low, if a 1 is not pressed then B3 will be high. Because pressing a 1 connects B0 and B3.

Similarly if 4 is pressed B4 will be low if not B4 will be high.

If 7 is pressed B5 will be low if not B5 will be high.
If * is pressed B6 will be low if not B6 will be high.

In other words when we set B0 low if any of the keys in column1 are pressed then the corresponding input to the microcontroller will go low and the program will output the binary number equivalent of the key that has been pressed.

If none of the keys in column1 are pressed then we move onto column2.

The code for scanning column1 is as follows:

These 3 lines set up PORTB with B0 = 0, B1 = 1 and B2 = 1.

```
COLUMN1    BCF         PORTB,0      ;Clear B0
           BSF         PORTB,1      ;Set B1
           BSF         PORTB,2      ;Set B2
```

These next 4 lines test input B3 to see if it clear if it is then a 1 is placed on PORTA, then the program continues. If B3 is set then we proceed to check to see if key 4 has been pressed, with CHECK4.

```
CHECK1     BTFSC       PORTB,3      ;Is B3 Clear?
           GOTO        CHECK4       ;No
           MOVLW       .1           ;Yes, output 1
           MOVWF       PORTA        ;to PORTA
```

These next 4 lines test input B4 to see if it clear if it is then a 4 is placed on PORTA, then the program continues. If B4 is set then we proceed to check to see if key 7 has been pressed, with CHECK7.

```
CHECK4     BTFSC       PORTB,4      ;Is B4 Clear?
           GOTO        CHECK7       ;No
           MOVLW       .4           ;Yes, output 4.
           MOVWF       PORTA
```

These next 4 lines test input B5 to see if it clear if it is then a 7 is placed on PORTA, then the program continues. If B5 is set then we proceed to Check to see if key * has been pressed, with CHECK11.

```
CHECK7     BTFSC       PORTB,5      ;Is B5 Clear?
           GOTO        CHECK11      ;No
           MOVLW       .7           ;Yes, output 7.
           MOVWF       PORTA
```

These next 4 lines test input B6 to see if it clear if it is then an 11 is placed on PORTA, then the program continues. If B5 is set then we proceed to check the keys in column2, with COLUMN2.

```
CHECK11    BTFSC        PORTB,6      ;Is B6 Clear?
           GOTO         COLUMN2      ;No
           MOVLW        .11          ;Yes, output 11.
           MOVWF        PORTA
```

These 3 lines set up PORTB with $B0 = 1$, $B1 = 0$ and $B2 = 1$.

```
COLUMN2  BSF          PORTB,0      ;Set B0
         BCF          PORTB,1      ;Clear B1
         BSF          PORTB,2      ;Set B2
```

We then check to see if key2 has been pressed by testing to see if B3 is clear, if it is then a 2 is placed on PORTA and the program continues. If B3 is set then we proceed with CHECK5. This code is:

```
CHECK2     BTFSC        PORTB,3      ;Is B3 Clear?
           GOTO         CHECK5       ;No
           MOVLW        .2           ;Yes, output 2.
           MOVWF        PORTA
```

The program continues in the same manner checking 5, 8 and 10 (0). Then moving onto column3 to check for 3, 6, 9 and 12 (#). After completing the scan the program then goes back to continue the scan again.

It takes about 45 lines of code to complete a scan of the keypad. With a 32,768Hz crystal the lines of code are executed at ¼ of this speed i.e. 8192 lines per second. So the scan time is $45/8192 = 5.5$ms. This is why no matter how quickly you press the key the microcontroller will be able to detect it.

Security code

Probably one of the most useful applications of a keypad is to enter a code to turn something on and off such as a burglar alarm or door entry system.

In the following program KEYS3.ASM the sub-routine SCAN, scans the keypad, waits for a key to be pressed, waits 0.1 seconds for the bouncing to stop, waits for the key to be released, waits 0.1 seconds for the bouncing to stop and then returns with the key number in W which can then be transferred into a file.

This is then used as a security code to turn on an LED (PORTA,0) when 3 digits (137) have been pressed and turn the LED off again when the same 3 digits are pressed. You can of course use any 3 digits.

;KEYS3.ASM

;EQUATES SECTION

```
ZEROBIT     EQU     2
TMR0        EQU     1
STATUS      EQU     3          ;means STATUS is file 3.
PORTA       EQU     5          ;means PORTA is file 5.
PORTB       EQU     6          ;means PORTB is file 6.
TRISA       EQU     85H
TRISB       EQU     86H
OPTION_R    EQU     81H
NUM1        EQU     0CH
NUM2        EQU     0DH
NUM3        EQU     0EH
;****************************************************
;
LIST        P=16F84            ;we are using the 16F84.
ORG         0                  ;the start address in memory is 0
GOTO        START              ;goto start!


;****************************************************
;
;SUB-ROUTINE SECTION

SCAN        NOP

COLUMN1     BCF     PORTB,0         ;Clear B0

            BSF     PORTB,1         ;Set B1
            BSF     PORTB,2         ;Set B2

CHECK1      BTFSC   PORTB,3         ;Is B3 Clear?
            GOTO    CHECK4          ;No
            CALL    DELAYP1
CHECK1A     BTFSS   PORTB,3
            GOTO    CHECK1A
            CALL    DELAYP1
            RETLW   .1

CHECK4      BTFSC   PORTB,4         ;Is B4 Clear?
            GOTO    CHECK7          ;No
            CALL    DELAYP1
```

```
CHECK4A     BTFSS       PORTB,4
            GOTO        CHECK4A
            CALL        DELAYP1
            RETLW       .4

CHECK7      BTFSC       PORTB,5         ;Is B5 Clear?
            GOTO        CHECK11         ;No
            CALL        DELAYP1
CHECK7A     BTFSS       PORTB,5
            GOTO        CHECK7A
            CALL        DELAYP1
            RETLW       .7

CHECK11     BTFSC       PORTB,6         ;Is B6 Clear?
            GOTO        COLUMN2         ;No
            CALL        DELAYP1
CHECK11A    BTFSS       PORTB,6
            GOTO        CHECK11A
            CALL        DELAYP1
            RETLW       .11

COLUMN2     BSF         PORTB,0         ;Set B0
            BCF         PORTB,1         ;Clear B1
            BSF         PORTB,2         ;Set B2

CHECK2      BTFSC       PORTB,3         ;Is B3 Clear?
            GOTO        CHECK5          ;No
            CALL        DELAYP1
CHECK2A     BTFSS       PORTB,3
            GOTO        CHECK2A
            CALL        DELAYP1
            RETLW       .2              ;Yes, output 2.

CHECK5      BTFSC       PORTB,4         ;Is B4 Clear?
            GOTO        CHECK8          ;No
            CALL        DELAYP1
CHECK5A     BTFSS       PORTB,4
            GOTO        CHECK5A
            CALL        DELAYP1
            RETLW       .5              ;Yes, output 5.

CHECK8      BTFSC       PORTB,5         ;Is B5 Clear?
            GOTO        CHECK0          ;No
            CALL        DELAYP1
CHECK8A     BTFSS       PORTB,5
            GOTO        CHECK8A
            CALL        DELAYP1
            RETLW       .8              ;Yes, output 8.
```

```
CHECK0       BTFSC    PORTB,6        ;Is B6 Clear?
             GOTO     COLUMN3        ;No
             CALL     DELAYP1
CHECK0A      BTFSS    PORTB,6
             GOTO     CHECK0A
             CALL     DELAYP1
             RETLW    0              ;Yes, output 10.
COLUMN3      BSF      PORTB,0        ;Set B0
             BSF      PORTB,1        ;Set B1
             BCF      PORTB,2        ;Clear B2

CHECK3       BTFSC    PORTB,3        ;Is B3 Clear?
             GOTO     CHECK6         ;No
             CALL     DELAYP1
CHECK3A      BTFSS    PORTB,3
             GOTO     CHECK3A
             CALL     DELAYP1
             RETLW    .3             ;Yes, output 3.

CHECK6       BTFSC    PORTB,4        ;Is B4 Clear?
             GOTO     CHECK9         ;No
             CALL     DELAYP1
CHECK6A      BTFSS    PORTB,4
             GOTO     CHECK6A
             CALL     DELAYP1
             RETLW    .6             ;Yes, output 6.

CHECK9       BTFSC    PORTB,5        ;Is B5 Clear?
             GOTO     CHECK12        ;No
             CALL     DELAYP1
CHECK9A      BTFSS    PORTB,5
             GOTO     CHECK9A
             CALL     DELAYP1
             RETLW    .9             ;Yes, output 9.

CHECK12      BTFSC    PORTB,6        ;Is B6 Clear?
             GOTO     COLUMN1        ;No
             CALL     DELAYP1
CHECK12A     BTFSS    PORTB,6
             GOTO     CHECK12A
             CALL     DELAYP1
             RETLW    .12            ;Yes, output 12.

;3/32 second delay.
DELAYP1      CLRF     TMR0           ;Start TMR0.
LOOPD        MOVF     TMR0,W         ;Read TMR0 into W.
             SUBLW    .3             ;TIME-3
```

```
                BTFSS         STATUS,ZEROBIT      ;Check TIME-W = 0
                GOTO          LOOPD               ;Time is not = 3.
                RETLW         0                   ;Time is 3, return.
;*********************************************************
;CONFIGURATION SECTION

START           BSF           STATUS,5            ;Turns to Bank1.
                MOVLW         B'00000000'         ;PORTA is OUTPUT
                MOVWF         TRISA
                MOVLW         B'11111000'
                MOVWF         TRISB               ;PORTB is mixed I/O.
                MOVLW         B'00000111'
                MOVWF         OPTION_R
                BCF           STATUS,5            ;Return to Bank0.
                CLRF          PORTA               ;Clears PortA.
                CLRF          PORTB               ;Clears PortB.
;*********************************************************
;Program starts now.
;Enter 3 digit code here
                MOVLW         1                   ;First digit
                MOVWF         NUM1
                MOVLW         3                   ;Second digit
                MOVWF         NUM2
                MOVLW         7                   ;Third digit
                MOVWF         NUM3

BEGIN           CALL          SCAN                ;Get 1st number
                SUBWF         NUM1,W
                BTFSS         STATUS,ZEROBIT      ;IS NUMBER = 1?
                GOTO          BEGIN               ;No

                CALL          SCAN                ;Get 2nd number
                SUBWF         NUM2,W
                BTFSS         STATUS,ZEROBIT      ;IS NUMBER = 3?
                GOTO          BEGIN               ;No

                CALL          SCAN                ;Get 3rd number.
                SUBWF         NUM3,W
                BTFSS         STATUS,ZEROBIT      ;IS NUMBER = 7?
                GOTO          BEGIN               ;No
                BSF           PORTA,0             ;Turn on LED, 137 entered

TURN_OFF        CALL          SCAN                ;Get 1st number again
                SUBWF         NUM1,W
                BTFSS         STATUS,ZEROBIT      ;IS NUMBER = 1?
                GOTO          TURN_OFF            ;No
                CALL          SCAN                ;Get 2nd number
```

```
              SUBWF      NUM2,W
              BTFSS      STATUS,ZEROBIT   ;IS NUMBER = 3?
              GOTO       TURN_OFF         ;No
              CALL       SCAN             ;Get 3rd number.
              SUBWF      NUM3,W
              BTFSS      STATUS,ZEROBIT   ;IS NUMBER = 7?
              GOTO       TURN_OFF         ;No
              BCF        PORTA,0          ;Turn off LED.
              GOTO       BEGIN
END
```

How does the program work?

The ports are configured as in the previous code KEYPAD.ASM.

The KEYS3.ASM program looks for the first key press and then it compares the number pressed with the required number stored in a user file called NUM1. It then looks for the second key to be pressed. But because the microcontroller is so quick, the first number could be stored and the program looks for the second number, but our finger is still pressing the first number.

Anti-bounce routine

Also when a mechanical key is pressed or released it does not make or break cleanly, it bounces around. If the micro is allowed too, it is fast enough to see these bounces as key presses so we must slow it down.

- We look first of all for the switch to be pressed.
- Then wait 0.1 seconds for the switch to stop bouncing.
- We then wait for the switch to be released.
- We then wait 0.1 seconds for the bouncing to stop before continuing.

The switch has then been pressed and released indicating one action.
The 0.1 second delay is written in the Header as DELAYP1.

Scan routine

The scan routine used in KEYS3.ASM is written into the subroutine.

When called it waits for a key to be pressed and then returns with the number just pressed in W. It can be copied and used as a subroutine in any program using a keypad.

- The scan routine checks for key presses as in the previous example KEYPAD.ASM, Column1 checks for the numbers 1, 4, 7 and 11 being pressed in turn.

- If the 1 is not pressed then the routine goes on to check for a 4.
- If the 1 is pressed then the routine waits 0.1 second for the bouncing to stop.
- The program then waits for the key to be released.
- Waits again 0.1 seconds for the bouncing to stop,
- and then returns with a value of 1 in W.

Code for CHECK1:

```
CHECK1      BTFSC      PORTB,3      ;Is B3 Clear? Pressed?
            GOTO       CHECK4       ;No
            CALL       DELAYP1      ;Antibounce delay, B3 clear
CHECK1A     BTFSS      PORTB,3      ;Is B3 Set? Released?
            GOTO       CHECK1A      ;No
            CALL       DELAYP1      ;Antibounce delay, B3 Set
            RETLW      .1           ;Return with 1 in W.
```

If numbers 4, 7 or 11 are pressed the routine will return with the corresponding value in W.

If no numbers in column1 are pressed then the scan routine continues on to column2 and column3. If no keys are pressed then the routine loops back to the start of the scan routine to continue checking.

Storing the code

The code i.e. 137 is stored in the files NUM1, NUM2, NUM3 with the following code:

```
        MOVLW     1         ;First digit
        MOVWF     NUM1
        MOVLW     3         ;Second digit
        MOVWF     NUM2
        MOVLW     7         ;Third digit
        MOVWF     NUM3
```

Checking for the correct code

- We first of all CALL SCAN to collect the first digit, which returns with the number pressed in W.
- We then subtract the value of W from the first digit of our code stored in NUM1 with:

SUBWF NUM1,W.

This means SUBtract W from the File NUM1. The (,W) stores the result of the subtraction in W. Without (,W) the result would have been stored in NUM1 and the value changed!

- We then check to see if NUM1 and W are equal, i.e. a correct match. In this case the zerobit in the status register would be set. Indicating the result NUM1−W = zero. This is done with:

BTFSS STATUS,ZEROBIT

We skip and carry on if it is set, i.e. a match. If it isn't we return to BEGIN to scan again.

- With a correct first press we then carry on checking for a second and if correct a third press to match the correct code.
- When the correct code is pressed we turn on our LED with:

BSF PORTA,0

- We then run through a similar sequence and wait for the code to turn off the LED.

Notice that if you enter an incorrect digit you return to BEGIN or TURN_OFF. If you forget what key you have pressed then press an incorrect one and start again.

You could of course modify this program by adding a fourth digit to the program then turn on the LED. In which case you use another user file called NUM4. You could of course use a different code for switching off the output.

You can also beep a buzzer for half a second to give yourself an audible feedback that you had pressed a button.

As an extra security measure you could wait for a couple of seconds if an incorrect key had been pressed, or wait for 2 minutes if three wrong numbers had been entered.

The keypad routine opens up many different circuit applications.

The SCAN routine can be copied and then pasted into any program using the keypad. Then when you CALL SCAN the program will return with the number pressed in W for you to do with it as you wish.

8
Program examples

New instructions used in this chapter:

- INCF
- INCFSZ
- DECF
- ADDWF

Counting events

Counting of course is a useful feature for any control circuit. We may wish to count the number of times a door has opened or closed, or count a number of pulses from a rotating disc. If we count cars into a car park we would increment a file count every time a car entered, using the instruction INCF COUNT. If we needed to know how many cars were in the car park we would have course have to reduce the count by one every time a car left. We would do this by DECF COUNT. To clear the user file COUNT to start we would CLRF COUNT. In this way the file count would store the number of cars in the car park. If you prefer COUNT could be called CARS. It is a user file call it what you like.

Let's look at an application.

Design a circuit that will count 10 presses of a switch, then turn an LED on and reset when the next ten presses are started. The hardware is that of Figure 5.1 with A0 as the switch input and B0 as the output to the LED.

There are two ways to count, UP and DOWN. We usually count up and know automatically when we have reached 10. A computer however knows when it reaches a count of 10 by subtracting the count from 10. If the answer is zero, then bingo. A simpler way however is to start at 10 and count down to zero – after 10 events we will have reached zero without doing a subtraction. Zero for the microcontroller is a really useful number.

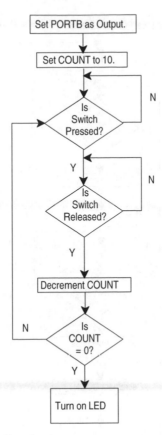

Figure 8.1 Initial counting flowchart

The initial flowchart for this problem is shown in Figure 8.1.

To ensure that the LED is OFF after the switch is pressed for the eleventh time put in TURN OFF LED after the switch is pressed, as shown in Figure 8.2.

N.B. The switch will bounce and the micro is fast enough to count these bounces, thinking that the switch has been pressed several times. A 0.1 second delay is inserted after each switch operation to allow time for the bounces to stop.

The final flowchart is shown in Figure 8.2.

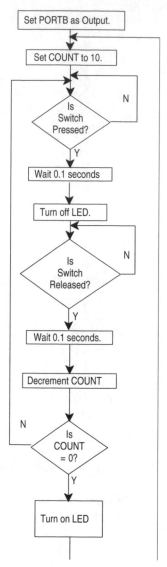

Figure 8.2 Final counting flowchart

The program for the counting circuit

;COUNT84.ASM using the 16F84 with a 32kHz. crystal

;EQUATES SECTION

```
TMR0        EQU     1           ;means TMR0 is file 1.
STATUS      EQU     3           ;means STATUS is file 3.
```

```
PORTA     EQU    5        ;means PORTA is file 5.
PORTB     EQU    6        ;means PORTB is file 6.
TRISA     EQU    85H      ;TRISA (the PORTA I/O selection) is
                          ;file 85H
TRISB     EQU    86H      ;TRISB (the PORTB I/O selection) is
                          ;file 86H
OPTION_R  EQU    81H      ;the OPTION register is file 81H
ZEROBIT   EQU    2        ;means ZEROBIT is bit 2.
COUNT     EQU    0CH      ;COUNT is file 0C, a register to count
                          ;events.
```

;**
```
LIST      P = 16F84      ;we are using the 16F84.
ORG       0              ;the start address in memory is 0
GOTO      START          ;goto start!
```

;**
;Configuration Bits

```
__CONFIG H'3FF0'   ;selects LP oscillator, WDT off, PUT on,
                   ;Code Protection disabled.
```

;**
;SUBROUTINE SECTION.

;3/32 second delay.
```
DELAY  CLRF    TMR0                    ;START TMR0.
LOOPA  MOVF    TMR0,W                  ;READ TMR0 INTO W.
       SUBLW   .3                      ;TIME - 3
       BTFSS   STATUS, ZEROBIT         ;Check TIME-W = 0
       GOTO    LOOPA                   ;Time is not = 3.
       RETLW   0                       ;Time is 3, return.
```

;**
;CONFIGURATION SECTION

```
START  BSF     STATUS,5        ;Turns to Bank1.

       MOVLW   B'00011111'     ;5bits of PORTA are I/P
       MOVWF   TRISA

       MOVLW   B'00000000'
       MOVWF   TRISB           ;PORTB is OUTPUT
```

```
                MOVLW       B'00000111'    ;Prescaler is /256
                MOVWF       OPTION_R       ;TIMER is 1/32 secs.

                BCF         STATUS,5       ;Return to Bank0.
                CLRF        PORTA          ;Clears PortA.
                CLRF        PORTB          ;Clears PortB.

;**********************************************************
;
;Program starts now.
BEGIN           MOVLW       .10
                MOVWF       COUNT          ;Put 10 into COUNT.
PRESS           BTFSC       PORTA,0        ;Check switch is pressed
                GOTO        PRESS
                CALL        DELAY          ;Wait for 3/32 seconds.
                BCF         PORTB,0        ;TURN OFF LED.
RELEASE         BTFSS       PORTA,0        ;Check switch is released.
                GOTO        RELEASE
                CALL        DELAY          ;WAIT for 3/32 seconds.
                DECFSZ      COUNT          ;Dec COUNT skip if 0.
                GOTO        PRESS          ;Wait for another press.
                BSF         PORTB,0        ;Turn on LED.
                GOTO        BEGIN          ;Restart
END
```

How does it work?

- The file COUNT is first loaded with the count i.e. 10 with:

```
MOVLW           .10
MOVWF           COUNT          ;Put 10 into COUNT.
```

- We then wait for the switch to be pressed, by PORTA,0 going low:

```
PRESS    BTFSC     PORTA,0     ;Check switch is pressed
         GOTO      PRESS
```

- Anti-bounce:

```
CALL            DELAY          ;Wait for 3/32 seconds.
```

- Turn off the LED on B0:

```
BCF             PORTB,0
```

- Wait for switch to be released

```
RELEASE  BTFSS     PORTA,0     ;Check switch is released.
         GOTO      RELEASE
```

- Anti-bounce:

```
CALL   DELAY   ;Wait for 3/32 seconds.
```

- Decrement the file COUNT, if zero turn on LED and return to begin. If not zero continue pressing the switch.

DECFSZ	COUNT	;Dec COUNT skip if 0.
GOTO	PRESS	;Wait for another press.
BSF	PORTB,0	;Turn on LED.
GOTO	BEGIN	;Restart

This may appear to be a lot of programming to count presses of a switch, but once saved as a subroutine it can be reused in any other programs.

Look up table

A look up table is used to change data from one form to another i.e. pounds to kilograms, °C to °F, inches to centimeters etc. The explanation of the operation of a look up table is best understood by way of an example.

7-Segment display

Design a circuit that will count and display on a 7-segment display, the number of times a button is pressed, up to 10. The circuit diagram for this is shown in Figure 8.3.

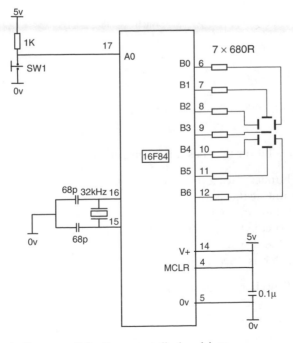

Figure 8.3 Circuit diagram of the 7-segment display driver

The flowchart for the 7-Segment Display Driver is shown in Figure 8.4.

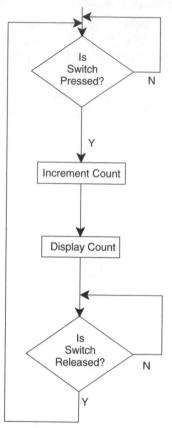

Figure 8.4 Initial flowchart for the 7-segment driver

This is a basic solution that has a few omissions:

- The switch bounces when pressed.
- Clear the count at the start.
- The micro counts in binary, we require a 7-segment decimal display. So we need to convert the binary count to drive the relevant segments on the display.
- When the switch is released it bounces.

The amended flowchart is shown in Figure 8.5.

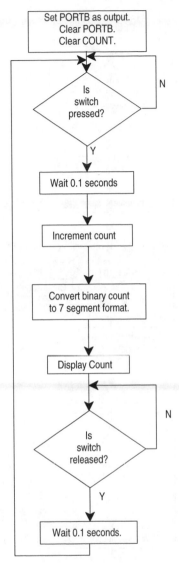

Figure 8.5 Amended flowchart for 7-segment display

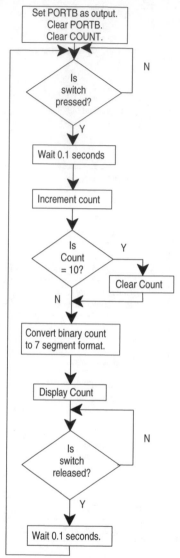

Figure 8.6 Final flowchart for 7-segment display

The flowchart is missing just one thing! What happens when the count reaches 10? The counter needs resetting (it would count up to 255 before resetting). The final flowchart is shown in Figure 8.6.

Now about this look up table:

Table 8.1 shows the configuration of **PORTB** to drive the 7-segment display. (Refer also to Figure 8.3).

Table 8.1 Binary code to drive 7-segment display

NUMBER	PORTB							
	B7	B6	B5	B4	B3	B2	B1	B0
0	0	1	1	1	0	1	1	1
1	0	1	0	0	0	0	0	1
2	0	0	1	1	1	0	1	1
3	0	1	1	0	1	0	1	1
4	0	1	0	0	1	1	0	1
5	0	1	1	0	1	1	1	0
6	0	1	1	1	1	1	0	0
7	0	1	0	0	0	0	1	1
8	0	1	1	1	1	1	1	1
9	0	1	0	0	1	1	1	1

The look up table for this is:

```
CONVERT    ADDWF      PC
           RETLW      B'01110111'    ;0
           RETLW      B'01000001'    ;1
           RETLW      B'00111011'    ;2
           RETLW      B'01101011'    ;3
           RETLW      B'01001101'    ;4
           RETLW      B'01101110'    ;5
           RETLW      B'01111100'    ;6
           RETLW      B'01000011'    ;7
           RETLW      B'01111111'    ;8
           RETLW      B'01001111'    ;9
```

How does the look up table work?

Suppose we need to display a 0.

We move 0 into W and CALL the look up table, here it is called CONVERT.

The first line says ADD W to the Program Count, since $W = 0$ then go to the next line of the program which will return with the 7-segment value 0.

Suppose we need to display a 6.

Move 6 into W and CALL CONVERT. The first line says ADD W to the Program Count, since W contains 6 then go to the next line of the program and move down 6 more lines and return with the code for 6, etc.

Just one more thing: To check that a count has reached 10, subtract 10 from the count if the answer is 0, bingo!

The program listing for the complete program is:

;DISPLAY.ASM

;EQUATES SECTION

```
PC          EQU    2        ;means PC is file 2.
TMR0        EQU    1        ;means TMR0 is file 1.
STATUS      EQU    3        ;means STATUS is file 3.
PORTA       EQU    5        ;means PORTA is file 5.
PORTB       EQU    6        ;means PORTB is file 6.
TRISA       EQU    85H      ;TRISA (the PORTA I/O selection) is file 85H
TRISB       EQU    86H      ;TRISB (the PORTB I/O selection) is file 86H
OPTION_R    EQU    81H      ;the OPTION register is file 81H
ZEROBIT     EQU    2        ;means ZEROBIT is bit 2.
COUNT       EQU    0CH      ;COUNT is file 0C, a register to count events.
```

;***

```
LIST    P = 16F84    ;we are using the 16F84.
ORG     0            ;the start address in memory is 0
GOTO    START        ;goto start!
```

;***
;Configuration Bits
__CONFIG H'3FF0'   ;selects LP oscillator, WDT off, PUT on,
                     ;Code Protection disabled.

;***
;SUBROUTINE SECTION.

;3/32 second delay.
```
DELAY       CLRF     TMR0                 ;START TMR0.
LOOPA       MOVF     TMR0,W               ;READ TMR0 INTO W.
            SUBLW    .3                   ;TIME - 3
            BTFSS    STATUS, ZEROBIT      ;Check TIME-W = 0
            GOTO     LOOPA                ;Time is not = 3.
            RETLW    0                    ;Time is 3, return.

CONVERT     ADDWF    PC
            RETLW    B'01110111'          ;0
            RETLW    B'01000001'          ;1
            RETLW    B'00111011'          ;2
            RETLW    B'01101011'          ;3
            RETLW    B'01001101'          ;4
            RETLW    B'01101110'          ;5
```

```
                    RETLW      B'01111100'      ;6
                    RETLW      B'01000011'      ;7
                    RETLW      B'01111111'      ;8
                    RETLW      B'01001111'      ;9
```

;***
;CONFIGURATION SECTION

```
START    BSF          STATUS,5          ;Turns to Bank1.

         MOVLW        B'00011111'       ;5bits of PORTA are I/P
         MOVWF        TRISA

         MOVLW        B'00000000'
         MOVWF        TRISB             ;PORTB is OUTPUT

         MOVLW        B'00000111'       ;Prescaler is /256
         MOVWF        OPTION_R          ;TIMER is 1/32 secs.

         BCF          STATUS,5          ;Return to Bank0.
         CLRF         PORTA             ;Clears PortA.
         CLRF         PORTB             ;Clears PortB.
```

;***
;Program starts now.

```
             CLRF      COUNT                  ;Set COUNT to 0.
PRESS        BTFSC     PORTA,0                ;Test for switch press.
             GOTO      PRESS                  ;Not pressed.
             CALL      DELAY                  ;Antibounce wait 0.1sec.
             INCF      COUNT                  ;Add 1 to COUNT.
             MOVF      COUNT,W                ;Move COUNT to W.
             SUBLW     .10                    ;COUNT-10, W is altered.
             BTFSC     STATUS,ZEROBIT         ;Is COUNT - 10 = 0?
             CLRF      COUNT                  ;Count = 10 Make Count = 0
             MOVF      COUNT,W                ;Put Count in W again.
             CALL      CONVERT                ;Count is not 10, carry on.
             MOVWF     PORTB                  ;Output number to display.

RELEASE      BTFSS     PORTA,0                ;Is switch released?
             GOTO      RELEASE                ;Not released.
             CALL      DELAY                  ;Antibounce wait 0.1sec.
             GOTO      PRESS                  ;Look for another press.
END
```

How does the program work?

- The file count is cleared (to zero) and we wait for the switch to be pressed.

```
         CLRF         COUNT          ;Set COUNT to 0.
PRESS    BTFSC        PORTA,0        ;Test for switch press.
         GOTO         PRESS          ;Not pressed.
```

- Wait for 0.1 seconds, Anti-bounce.

```
CALL                 DELAY
```

- Add 1 to COUNT and check to see if it 10:

```
         INCF         COUNT          ;Add 1 to COUNT.
         MOVF         COUNT,W        ;Move COUNT to W.
         SUBLW        .10            ;COUNT-10, W is altered.
         BTFSC        STATUS,ZEROBIT ;Is COUNT - 10 = 0?
```

- If COUNT is 10, Clear it to 0 and output the count as 0. If the COUNT is not 10 then output the count.

```
         CLRF         COUNT          ;Count = 10 Make Count = 0
         MOVF         COUNT,W        ;Put Count in W again.
         CALL         CONVERT        ;Count is not 10, carry on.
         MOVWF        PORTB          ;Output number to display.
```

- Wait for the switch to be released and de-bounce. Then return to monitor the presses.

```
RELEASE  BTFSS        PORTA,0        ;Is switch released?
         GOTO         RELEASE        ;Not released.
         CALL         DELAY          ;Antibounce wait 0.1sec.
         GOTO         PRESS          ;Look for another press.
```

Test your understanding

- Modify the program to Count up to 6 and reset.
- Modify the program to Count up to F in HEX and reset.

A look up table to change °C to °F is shown below, called DEGREE

```
DEGREE   ADDWF        PC             ;ADD W to Program Count.
         RETLW        .32            ;0°C  = 32°F
         RETLW        .34            ;1°C = 34°F
         RETLW        .36            ;2°C = 36°F
         RETLW        .37            ;3°C = 37°F
```

RETLW	.39	;4°C = 39°F
RETLW	.41	;5°C = 41°F
RETLW	.43	;6°C = 43°F
RETLW	.45	;7°C = 45°F
RETLW	.46	;8°C = 46°F
RETLW	.48	;9°C = 48°F
RETLW	.50	;10°C = 50°F
RETLW	.52	;11°C = 52°F
RETLW	.54	;12°C = 54°F
RETLW	.55	;13°C = 55°F
RETLW	.57	;14°C = 57°F
RETLW	.59	;15°C = 59°F
RETLW	.61	;16°C = 61°F
RETLW	.63	;17°C = 63°F
RETLW	.64	;18°C = 64°F
RETLW	.66	;19°C = 66°F
RETLW	.68	;20°C = 68°F
RETLW	.70	;21°C = 70°F
RETLW	.72	;22°C = 72°F
RETLW	.73	;23°C = 73°F
RETLW	.75	;24°C = 75°F
RETLW	.77	;25°C = 77°F
RETLW	.79	;26°C = 79°F
RETLW	.81	;27°C – 81°F
RETLW	.82	;28°C = 82°F
RETLW	.84	;29°C = 84°F
RETLW	.86	;30°C = 86°F

Another application of the use of the look up table is a solution for a previous example i.e. the "Control Application – A Hot Air Blower." Introduced in Chapter 5.

In this example when **PORTA** was read the data was treated as a binary number, but we could just as easily treat the data as decimal number.

i.e. A2 A1 A0 = 000 or 0
$\qquad$ = 001 or 1
$\qquad$ = 010 or 2
$\qquad$ = 011 or 3
$\qquad$ = 100 or 4
$\qquad$ = 101 or 5
$\qquad$ = 110 or 6
$\qquad$ = 111 or 7

The look up table for this would be:

```
CONVERT    ADDWF    PC
           RETLW    B'00000010'    ;0 on PORTA turns on B1
           RETLW    B'00000001'    ;1 on PORTA turns on B0
           RETLW    B'00000011'    ;2 on PORTA turns on B1,B0
           RETLW    B'00000001'    ;3 on PORTA turns on B0
           RETLW    B'00000000'    ;4 on PORTA turns off B1,B0
           RETLW    B'00000001'    ;5 on PORTA turns on B0
           RETLW    B'00000000'    ;6 on PORTA turns off B1,B0
           RETLW    B'00000010'    ;7 on PORTA turns on B1
```

The complete program listing for the program DISPLAY2 would be:

;DISPLAY2.ASM

;EQUATES SECTION

```
PC         EQU    2        ;Program Counter is file 2.
TMR0       EQU    1        ;means TMR0 is file 1.
STATUS     EQU    3        ;means STATUS is file 3.
PORTA      EQU    5        ;means PORTA is file 5.
PORTB      EQU    6        ;means PORTB is file 6.
TRISA      EQU    85H      ;TRISA (the PORTA I/O selection) is
                          ;file 85H
TRISB      EQU    86H      ;TRISB (the PORTB I/O selection) is
                          ;file 86H
OPTION_R   EQU    81H      ;the OPTION register is file 81H
ZEROBIT    EQU    2        ;means ZEROBIT is bit 2.
COUNT      EQU    0CH      ;COUNT is file 0C, a register to count
                          ;events.
```

```
;****************************************************
LIST    P = 16F84    ;we are using the 16F84.
ORG     0            ;the start address in memory is 0
GOTO    START        ;goto start!

;****************************************************
;Configuration Bits

__CONFIG H'3FF0'    ;selects LP oscillator, WDT off, PUT on,
                    ;Code Protection disabled.

;****************************************************
```

;SUBROUTINE SECTION.

```
CONVERT    ADDWF      PC
           RETLW      B'00000010'    ;0 on PORTA turns on B1
           RETLW      B'00000001'    ;1 on PORTA turns on B0
           RETLW      B'00000011'    ;2 on PORTA turns on B1,B0
           RETLW      B'00000001'    ;3 on PORTA turns on B0
           RETLW      B'00000000'    ;4 on PORTA turns off B1,B0
           RETLW      B'00000001'    ;5 on PORTA turns on B0
           RETLW      B'00000000'    ;6 on PORTA turns off B1,B0
           RETLW      B'00000010'    ;7 on PORTA turns on B1
```

;**

;CONFIGURATION SECTION

```
START   BSF        STATUS,5       ;Turns to Bank1.

        MOVLW      B'00011111'    ;5bits of PORTA are I/P
        MOVWF      TRISA

        MOVLW      B'00000000'
        MOVWF      TRISB          ;PORTB is OUTPUT

        MOVLW      B'00000111'    ;Prescaler is /256
        MOVWF      OPTION_R       ;TIMER is 1/32 secs.

        BCF        STATUS,5       ;Return to Bank0.
        CLRF       PORTA          ;Clears PortA.
        CLRF       PORTB          ;Clears PortB.
```

;**

;Program starts now.

```
BEGIN   MOVF       PORTA,W    ;Read PORTA into W
        CALL       CONVERT    ;Obtain O/Ps from I/Ps.
        MOVWF      PORTB      ;switch on O/Ps
        GOTO       BEGIN      ;repeat

END
```

How does the program work?

- The program first of all reads the value of PORTA into the working register, W:

```
        MOVF       PORTA,W
```

- The CONVERT routine is called which returns with the correct setting of the outputs in W. i.e. If the value of PORTA was 3 then the look up table would return with 00000001 in W to turn on B0 and turn off B1:

> CALL   CONVERT ;Obtain O/Ps from I/Ps.
> MOVWF   PORTB   ;switch on O/Ps

- The program then returns to check the setting of PORTA again.

Numbers larger than 255

The PIC Microcontrollers are 8 bit devices, this means that they can easily count up to 255 using one memory location. But to count higher then more than one memory location has to be used for the count.

Consider counting a switch press up to 1000 and then turn on an LED to show this count has been achieved. The circuit for this is shown in Figure 8.7.

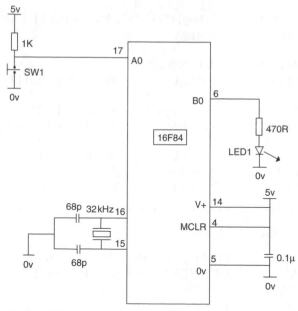

Figure 8.7 Circuit for 1000 count

To count up to 1000 in decimal i.e. 03E8 in hex, files COUNTB and COUNTA will store the count (a count of 65535 is then possible).

COUNTB will count up to 03H then when COUNTA has reached E8H, LED1 will light indicating the count of 1000 has been reached.

The flowchart for this 1000 count is shown in Figure 8.8.

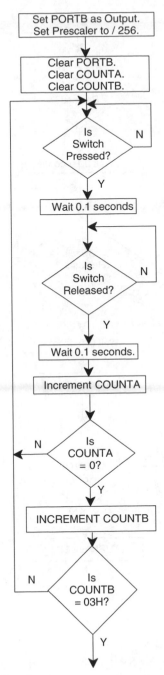

Figure 8.8 Count of 1000 flowchart

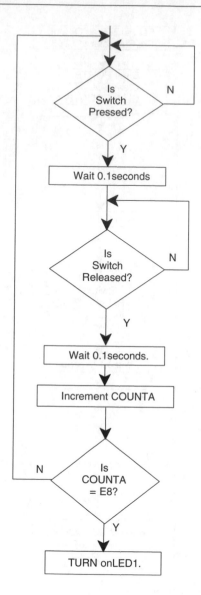

Figure 8.8 Continued

Flowchart explanation

- The program is waiting for SW1 to be pressed. When it is, there is a delay of 0.1 seconds to allow the switch bounce to stop.
- The program then looks for the switch to be released and waits 0.1 seconds for the bounce to stop.

- 1 is then added to COUNTA and a check is made to see if the count has overflowed i.e. reached 256. (255 is the maximum it will hold, when it reaches 256 it will reset to zero just like a two digit counter would reset to zero going from 99 to 100.)
- If COUNTA has overflowed then we increment COUNTB.
- A check is made to see if COUNTB has reached 03H, if not we return to keep counting.
- If COUNTB has reached 03H then we count presses until COUNTA reaches E8H. The count in decimal is then 1000 and the LED is lit.

Any count can be attained by altering the values COUNTB and COUNTA are allowed to count up to i.e. to count up to 5000 in decimal which is 1388H. Ask if COUNTB = 13H then count until COUNTA has reached 88H.

The program listing

;CNT1000.ASM

;EQUATES SECTION

```
TMR0        EQU   1        ;means TMR0 is file 1.
STATUS      EQU   3        ;means STATUS is file 3.
PORTA       EQU   5        ;means PORTA is file 5.
PORTB       EQU   6        ;means PORTB is file 6.
TRISA       EQU   85H      ;TRISA (the PORTA
                           ;I/O selection) is file 85H
TRISB       EQU   86H      ;TRISB (the PORTB I/O selection) is file 86H
OPTION_R    EQU   81H      ;the OPTION register is file 81H
ZEROBIT     EQU   2        ;means ZEROBIT is bit 2.
COUNTA      EQU   0CH      ;USER RAM LOCATION.
COUNTB      EQU   0DH      ;USER RAM LOCATION.
```

```
;********************************************************
;
LIST        P = 16F84     ;we are using the 16F84.
ORG         0             ;the start address in memory is 0
GOTO        START         ;goto start!
```

```
;********************************************************
;
;Configuration Bits

__CONFIG H'3FF0'          ;selects LP oscillator, WDT off, PUT on,
                          ;Code Protection disabled.
```

```
;********************************************************
;
```

;SUBROUTINE SECTION.

;3/32 second delay.

```
DELAY    CLRF     TMR0              ;START TMR0.
LOOPA    MOVF     TMR0,W            ;READ TMR0 INTO W.
         SUBLW    .3                ;TIME - 3
         BTFSS    STATUS,ZEROBIT    ;Check TIME-W = 0
         GOTO     LOOPA             ;Time is not = 3.
         RETLW    0                 ;Time is 3, return.
```

;***

;CONFIGURATION SECTION

```
START    BSF      STATUS,5          ;Turns to Bank1.

         MOVLW    B'00011111'       ;5bits of PORTA are I/P
         MOVWF    TRISA

         MOVLW    B'00000000'
         MOVWF    TRISB             ;PORTB is OUTPUT

         MOVLW    B'00000111'       ;Prescaler is /256
         MOVWF    OPTION_R          ;TIMER is 1/32 secs.

         BCF      STATUS,5          ;Return to Bank0.
         CLRF     PORTA             ;Clears PortA.
         CLRF     PORTB             ;Clears PortB.
```

;***
;Program starts now.

```
         CLRF     COUNTA
         CLRF     COUNTB

PRESS    BTFSC    PORTA,0           ;Check switch pressed
         GOTO     PRESS
         CALL     DELAY             ;Wait for 3/32 seconds.
RELEASE  BTFSS    PORTA,0           ;Check switch is released.
         GOTO     RELEASE
         CALL     DELAY             ;Wait for 3/32 seconds.
```

```
            INCFSZ   COUNTA              ;Inc. COUNT skip if 0.
            GOTO     PRESS
            INCF     COUNTB
            MOVLW    03H                 ;Put 03H in W.              *
            SUBWF    COUNTB,W            ;COUNTB - W (i.e. 03)
            BTFSS    STATUS,ZEROBIT      ;IS COUNTB = 03H
            GOTO     PRESS               ;No

PRESS1      BTFSC    PORTA,0             ;Check switch pressed.
            GOTO     PRESS1
            CALL     DELAY               ;Wait for 3/32 seconds.
RELEASE1    BTFSS    PORTA,0             ;Check switch released.
            GOTO     RELEASE1
            CALL     DELAY               ;Wait for 3/32 seconds.
            INCF     COUNTA
            MOVLW    0E8H                ;Put E8 in W.               *
            SUBWF    COUNTA              ;COUNTA - E8.
            BTFSS    STATUS,ZEROBIT      ;COUNTA = E8?
            GOTO     PRESS1              ;No.
            BSF      PORTB,0             ;Yes, turn on LED1.
STOP        GOTO     STOP                ;stop here

END
```

How does the program work?

- The two files used for counting are cleared.

```
            CLRF     COUNTA
            CLRF     COUNTB
```

- As we have done previously we wait for the switch to be pressed and released and to stop bouncing:

```
PRESS       BTFSC    PORTA,0      ;Check switch pressed
            GOTO     PRESS
            CALL     DELAY        ;Wait for 3/32 seconds.

RELEASE     BTFSS    PORTA,0      ;Check switch is released.
            GOTO     RELEASE
            CALL     DELAY        ;Wait for 3/32 seconds.
```

- We add1 to file COUNTA and check to see if it zero. If it isn't then continue monitoring presses. (The file would be zero when we add 1 to the 8 bit number 1111 1111, it overflows to 0000 0000):

```
INCFSZ    COUNTA    ;Inc. COUNT skip if 0.
GOTO      PRESS
```

- If the file COUNTA has overflowed then we add 1 to the file COUNTB, just like you would do with two columns of numbers. We then need to know if COUNTB has reached 03H. If COUNTB is not 03H then we return to PRESS and continue monitoring the presses.

```
INCF     COUNTB
MOVLW    03H              ;Put 03H in W.
SUBWF    COUNTB,W         ;COUNTB - W (i.e. 03)
BTFSS    STATUS,ZEROBIT   ;IS COUNTB = 03H?
GOTO     PRESS            ;No
```

- Once COUNTB has reached 03H we need only wait until COUNTA reaches 0E8H and we would have counted up to 03E8H i.e. 5000 in decimal. Then we turn on the LED.

```
PRESS1     BTFSC    PORTA,0          ;Check switch pressed.
           GOTO     PRESS1
           CALL     DELAY            ;Wait for 3/32 seconds.
RELEASE1   BTFSS    PORTA,0          ;Check switch released.
           GOTO     RELEASE1
           CALL     DELAY            ;Wait for 3/32 seconds.
           INCF     COUNTA
           MOVLW    0E8H             ;Put E8 in W.
           SUBWF    COUNTA           ;COUNTA – E8.
           BTFSS    STATUS,ZEROBIT   ;COUNTA = E8?
           GOTO     PRESS1           ;No.
           BSF      PORTB,0          ;Yes, turn on LED1.
STOP       GOTO     STOP             ;stop here
```

This listing can be used as a subroutine in your program to count up to any number to 65535 (or more if you use a COUNTC file). Just alter COUNTB and COUNTA values to whatever values you wish, in the two places marked * in the program.

Question. How would you count up to 20,000?

Answer. (Have you tried it first!!).
20,000 = 4E20H so COUNTB would count up to 4EH and COUNTA would then count to 20H.

Question. How would you count to 100,000?

Answer. 100,000 = 0186A0H, you would use a third file COUNTC to count to 01H, COUNTB would count to 86H and COUNTA would count to A0H.

Programming can be made a lot simpler by keeping a library of subroutines. Here is another . . . .

Long time intervals

Probably the more frequent use of a large count is to count TMR0 pulses to generate long time intervals. We have previously seen in the section on delay that we can slow the internal timer clock down to 1/32 seconds. Counting a maximum of 255 of these gives a time of $255 \times 1/32 = 8$ seconds. Suppose we want to turn on an LED for 5 minutes when a switch is pressed.

5 minutes = 300 seconds = 300×32 (1/32 seconds) i.e. a TMR0 count of 9600. This is 2580 in hex. The circuit is the same as Figure 8.7 for the 1000-count circuit, and the flowchart is shown in Figure 8.9.

Explanation of the flowchart

1. Wait until the switch is pressed, the LED is then turned on.
2. TMR0 is cleared to start the timing interval.
3. TMR0 is moved into W (read) to catch the first count.
4. Then wait for TMR0 to return to zero, (the count will be 256) i.e. 100 in hex.
5. COUNTA is then incremented and steps 3 and 4 repeated until COUNTA reaches 25H.
6. Wait until TMR0 has reached 80H.
7. The count has reached 2580H i.e. 9600 in decimal. 5 minutes has elapsed and the LED is turned off.

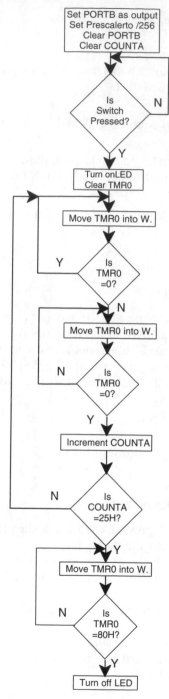

Figure 8.9 Flowchart for the 5 minute delay

Program listing for 5 minute delay

;LONGDLY.ASM

;EQUATES SECTION

```
TMR0       EQU   1      ;means TMR0 is file 1.
STATUS     EQU   3      ;means STATUS is file 3.
PORTA      EQU   5      ;means PORTA is file 5.
PORTB      EQU   6      ;means PORTB is file 6.
TRISA      EQU   85H    ;TRISA (the PORTA I/O selection) is file 85H
TRISB      EQU   86H    ;TRISB (the PORTB I/O selection) is file 86H
OPTION_R   EQU   81H    ;the OPTION register is file 81H
ZEROBIT    EQU   2      ;means ZEROBIT is bit 2.
COUNTA     EQU   0CH    ;COUNT is file 0C, a register to count events.
```

;**

```
LIST      P = 16F84    ;we are using the 16F84.
ORG       0            ;the start address in memory is 0
GOTO      START        ;goto start!
```

;**

;Configuration Bits

```
__CONFIG H'3FF0'      ;selects LP oscillator, WDT off, PUT on,
                      ;Code Protection disabled.
```

;**

;CONFIGURATION SECTION

```
START   BSF        STATUS,5     ;Turns to Bank1.

        MOVLW      B'00011111'  ;5bits of PORTA are I/P
        MOVWF      TRISA

        MOVLW      B'00000000'
        MOVWF      TRISB        ;PORTB is OUTPUT

        MOVLW      B'00000111'  ;Prescaler is /256
        MOVWF      OPTION_R     ;TIMER is 1/32 secs.

        BCF        STATUS,5     ;Return to Bank0.
        CLRF       PORTA        ;Clears PortA.
        CLRF       PORTB        ;Clears PortB.
```

;**

;Program starts now.

```
        CLRF       COUNTA
PRESS   BTFSC      PORTA,0                  ;Check switch pressed.
```

```
              GOTO      PRESS              ;No
              BSF       PORTB,0            ;Yes, turn on LED
              CLRF      TMR0               ;Start TMR0.
WAIT0         MOVF      TMR0,W             ;Move TMR0 into W
              BTFSC     STATUS,ZEROBIT     ;Is TMR0 = 0.
              GOTO      WAIT0              ;Yes
WAIT1         MOVF      TMR0,W             ;No, move TMR0 into W.
              BTFSS     STATUS,ZEROBIT
              GOTO      WAIT1              ;Wait for TMR0 to overflow
              INCF      COUNTA             ;Increment COUNTA
              MOVLW     25H
              SUBWF     COUNTA,W           ;COUNTA - 25H
              BTFSS     STATUS,ZEROBIT     ;Is COUNTA = 25H
              GOTO      WAIT0              ;COUNTA < 25H
WAIT2         MOVF      TMR0,W             ;COUNTA = 25H
              MOVLW     80H
              SUBWF     TMR0,W             ;TMR0 - 80H
              BTFSS     STATUS,ZEROBIT     ;Is TMR0 = 80H
              GOTO      WAIT2              ;TMR0 < 80H
              BCF       PORTB,0            ;TMR0 = 80H, turn off LED
END
```

The explanation of this program operation is similar to that of the count to 1000, done earlier in this chapter.

This listing can be used as a subroutine and times upto 65535 × 1/32 seconds i.e. 34 minutes can be obtained.

Problem: Change the listing to produce a 30 minute delay.

Hint. 1800sec in hex is 0708H.

One hour delay

Another and probably a simpler way of obtaining a delay of say 1 hour, is

- write a delay of 5 seconds,
- CALL it 6 times, this gives a delay of 30 seconds,
- put this in a loop to repeat 120 times, i.e.120 × 30 seconds = 1 hour.

This code for the 1 hour subroutine will look like:-

```
ONEHOUR       MOVLW     .120               ;put 120 in W
              MOVWF     COUNT              ;load COUNT with 120
LOOP          CALL      DELAY5             ;Wait 5 seconds
              CALL      DELAY5             ;Wait 5 seconds
```

```
CALL      DELAY5    ;Wait 5 seconds
CALL      DELAY5    ;Wait 5 seconds
CALL      DELAY5    ;Wait 5 seconds
CALL      DELAY5    ;Wait 5 seconds
DECFSZ    COUNT     ;Subtract 1 from COUNT
GOTO      LOOP      ;Count is not zero.
RETLW     0         ;RETURN to program.
```

The program for the one-hour delay

;ONEHOUR.ASM for 16F84. This sets PORTA as an INPUT (NB 1
;   means input) and PORTB as an OUTPUT
;   (NB 0 means output). The OPTION
;   register is set to /256 to give timing pulses
;   of 1/32 of a second.
;   1hour and 5 second delays are
;   included in the subroutine section.

;***
;EQUATES SECTION

```
TMR0       EQU   1      ;means TMR0 is file 1.
STATUS     EQU   3      ;means STATUS is file 3.
PORTA      EQU   5      ;means PORTA is file 5.
PORTB      EQU   6      ;means PORTB is file 6.
TRISA      EQU   85H    ;TRISA (the PORTA I/O selection) is file 85H
TRISB      EQU   86H    ;TRISB (the PORTB I/O selection) is file 86H
OPTION_R   EQU   81H    ;the OPTION register is file 81H
ZEROBIT    EQU   2      ;means ZEROBIT is bit 2.
COUNT      EQU   0CH    ;COUNT is file 0C, a register to count events.
```

;***
```
LIST    P = 16F84     ;we are using the 16F84.
ORG     0             ;the start address in memory is 0
GOTO    START         ;goto start!
```

;***
;Configuration Bits

```
__CONFIG H'3FF0'      ;selects LP oscillator, WDT off, PUT on,
                      ;Code Protection disabled.
```

;***
;SUBROUTINE SECTION.

;1 hour delay.
```
ONEHOUR              MOVLW          .120           ;put 120 in W
```

```
              MOVWF    COUNT              ;load COUNT with 120
LOOP          CALL     DELAY5             ;Wait 5 seconds
              CALL     DELAY5             ;Wait 5 seconds
              CALL     DELAY5             ;Wait 5 seconds
              CALL     DELAY5             ;Wait 5 seconds
              CALL     DELAY5             ;Wait 5 seconds
              CALL     DELAY5             ;Wait 5 seconds
              DECFSZ   COUNT              ;Subtract 1 from COUNT
              GOTO     LOOP               ;Count is not zero.
              RETLW    0                  ;RETURN to program.

;5 second delay.
DELAY5        CLRF     TMR0               ;START TMR0.
LOOPB         MOVF     TMR0,W             ;READ TMR0 INTO W.
              SUBLW    .160               ;TIME - 160
              BTFSS    STATUS,ZEROBIT     ;Check TIME-W = 0
              GOTO     LOOPB              ;Time is not = 160.
              RETLW    0                  ;Time is 160, return.
```

;**
;CONFIGURATION SECTION

```
START   BSF          STATUS,5   ;Turns to Bank1.

        MOVLW        B'00011111'  ;5bits of PORTA are I/P
        MOVWF        TRISA

        MOVLW        B'00000000'
        MOVWF        TRISB        ;PORTB is OUTPUT

        MOVLW        B'00000111'  ;Prescaler is /256
        MOVWF        OPTION_R     ;TIMER is 1/32 secs.

        BCF          STATUS,5     ;Return to Bank0.
        CLRF         PORTA        ;Clears PortA.
        CLRF         PORTB        ;Clears PortB.
```

;**
;Program starts now.
```
        BSF          PORTB,0   ;Turn on B0
        CALL         ONEHOUR   ;Wait 1 Hour.
        BCF          PORTB,0   ;Turn off B0.
STOP    GOTO         STOP      ;STOP!

        END
```

9
The 16C54 microcontroller

The 16C54 is an example of a one time programmable (OTP) device.

The 16C54 device was brought out before the 16F84.

The main difference between them is that the 16C54 is not electrically erasable, it has to be erased by UV light for about 15 minutes.

The 16C54 JW version is UV erasable.

The 16C54LP is a one time (only) programmable (OTP), 32 kHz version.

You would use a 16C54 JW for development and then program a OTP device for your final circuit. The OTP device has to be selected for the correct oscillator i.e. LP for 32kHz crystal, XT for 4MHz, HS for 20MHz and R-C for an R-C network.

The header for use with the 16C54 is shown below.

Header for the 16C54

```
; HEADER54.ASM for 16C54. This sets PORTA as an INPUT (NB 1
;                              means input) and PORTB as an OUTPUT
;                              (NB 0 means output). The OPTION
;                              register is set to /256 to give timing pulses
;                              of 1/32 of a second.
;                              1 second and 0.5 second delays are
;                              included in the subroutine section.
;
;*********************************************************************

; EQUATES SECTION
TMR0        EQU    1        ;means TMR0 is file 1.
STATUS      EQU    3        ;means STATUS is file 3.
PORTA       EQU    5        ;means PORTA is file 5.
PORTB       EQU    6        ;means PORTB is file 6.
ZEROBIT     EQU    2        ;means ZEROBIT is bit 2.
COUNT       EQU    7        ;means COUNT is file 7,
```

```
                                    ;a register to count events
TIME            EQU    8            ;file8 where the time is stored.
;****************************************************************

LIST            P=16C54            ; we are using the 16C54.
ORG             01FFH              ;the start address in memory is 1FF at the
                                   ;end.
GOTO            START              ; goto start!
ORG             0
;****************************************************************
```

;SUBROUTINE SECTION.

```
; 1 second delay.
DELAY1   CLRF     TMR0                   ;START TMR0.
LOOPA    MOVLW    .32
         MOVWF    TIME                   ;Time = 32/32 secs.
         MOVF     TMR0,W                 ;Read TMR0 into W.
         SUBWF    TIME,W                 ;TIME - 32, result in W.
         BTFSS    STATUS,ZEROBIT         ;Check TIME-W = 0
         GOTO     LOOPA                  ;Time is not = 32.
         RETLW    0                      ;Time is 32, return.

; 0.5 second delay.

DELAYP5  CLRF     TMR0                   ;START TMR0.
LOOPB    MOVLW    .16
         MOVWF    TIME                   ;Time = 16/32 secs.
         MOVF     TMR0,W                 ;READ TMR0 INTO W.
         SUBWF    TIME,W                 ;TIME - 16
         BTFSS    STATUS,ZEROBIT         ;Check TIME-W = 0
         GOTO     LOOPB                  ;Time is not = 16.
         RETLW    0                      ;Time is 16, return.
;****************************************************************
```

;CONFIGURATION SECTION

```
START        MOVLW    B'00001111'        ;4 bits of PORTA are I/P
             TRIS     PORTA
             MOVLW    B'00000000'
             TRIS     PORTB              ;PORTB is OUTPUT
             MOVLW    B'00000111'        ;Prescaler is /256
```

```
            OPTION                        ;TIMER is 1/32 secs.
            CLRF         PORTA            ;Clears PortA.
            CLRF         PORTB            ;Clears PortB.
```
.**
,
;Program starts now.

This header can now be used to write programs for the 16C54 Microcontroller.

There are a number of differences between the 16F84 and the 16C54 that the header has taken care of, but be aware of the differences when writing your program.

- The 16C54 does not use Banks so there is no need to change from one to the other.
- There are only 7 Registers on the 16C54 (see 16C54 Memory Map Table 9.1). So the user files start at number 7. i.e. COUNT EQU 7, TIME EQU 8.
- The 16C54 does not have the instruction SUBLW. So in the DELAY subroutine the delay is moved into a file called TIME. (NB. TIME EQUATES TO 8) Then the delay in the file is subtracted from W, giving the same result as for the 16F84.
- Why bother using the 16C54? The reprogrammable 16C54 i.e. 16C54JW is more expensive than the 16F84. But the one time programmable (OTP) 16C54 i.e. 16C54/04P is cheaper. So when your design is final you can blow the program into the cheaper 16C54/04P. Why bother with the expensive 16C54JW and not the 16F84 for program development? I don't know! Only convenience – not having to change the program.
- The 16C54JW has to be erased under an ultra violet lamp for about 15 minutes – this is a bind if you are impatient, you may need a couple.
- Pin 3 is only a T0CKI pin it does not double as A4 like the 16F84 and must be pulled high if the T0CKI is not being used.

16C54 memory map

Table 9.1 16C54 memory map

FILE ADDRESS	FILENAME
00	INDIRECT ADDRESS
01	TMR0
02	PC
03	STATUS
04	FSR
05	PORTA
06	PORTB
07	USER FILE
08	USER FILE
09	USER FILE
0A	USER FILE
0B	USER FILE
0C	USER FILE
0D	USER FILE
0E	USER FILE
0F	USER FILE
10	USER FILE
11	USER FILE
12	USER FILE
13	USER FILE
14	USER FILE
15	USER FILE
16	USER FILE
17	USER FILE
18	USER FILE
19	USER FILE
1A	USER FILE
1B	USER FILE
1C	USER FILE
1D	USER FILE
1E	USER FILE
1F	USER FILE

10
Alpha numeric displays

Using an Alpha Numeric Display in a project can bring it alive. Instead of showing a number on a 7 segment display the Alpha Numeric Display could indicate 'The Temperature is 27°C'. Instructions can also be given on screen.

This section details the use of a 16 character by 2 line display, which incorporates an HITACHI HD44780 Liquid Crystal Display Controller Driver Chip. The HD44780 is an industry standard also used in displays other than Hitachi (fortunately). The chip is also used as a driver for other display configurations i.e. 16×1, 20×2, 20×4, 40×2 etc. It has an on board character generator ROM which can display 240 character patterns.

The circuit diagram connecting the Alpha Numeric Display to the 16F84 is shown in Figure 10.1. This configuration is for the HD44780 driver and can be used with any of the displays using this chip.

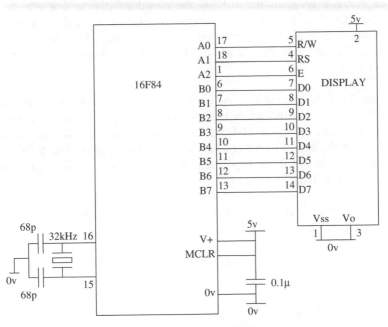

Figure 10.1 The 16F84 driving the alpha numeric display

Display pin identification

This display configuration shows 11 outputs from the Microcontroller, 3 control lines and 8 data lines connecting to the display.

R/W is the read/write control line, RS is the register select and E is the chip enable.

The R/W line tells the display to expect data to be written to it or to have data read from it. The data that is written to it is the address of the character, the code for the character or the type of command we require it to perform such as turn the cursor off.

The R/S line selects either a command to perform (R/S = 0) i.e. clear display, turn cursor on or off, or selects a data transfer (R/S = 1).

The E line enables, (E = 1) and disables, (E = 0) the display.

There is much more to this display than we are able to look at here. If you wish to know more about them you will need to consult the manufacturers data book.

If we use 11 lines to drive the display that would only leave 2 lines for the rest of our control with the 16F84. We could of course use a micro with 22 or 33 I/O. The display can however be driven with 4 data lines instead of 8, 4 bits of data are then sent twice. This complicates the program a little – but since I have done that work in the header it requires no more effort on your part.

Also the R/W line is used to write commands to the micro and read the busy line which indicates when the relatively slow display has processed the data. If we allow the micro enough time to complete its task then we do not have to read the busy line we can just write to the display. The R/W line can then be connected to 0v in a permanent write mode and we do not require a read/write line from the micro.

We will therefore only require 4 data lines and 2 control lines to drive the display leaving 7 lines available for I/O on the 16F84.

This 6 line control for the display is shown in Figure 10.2.

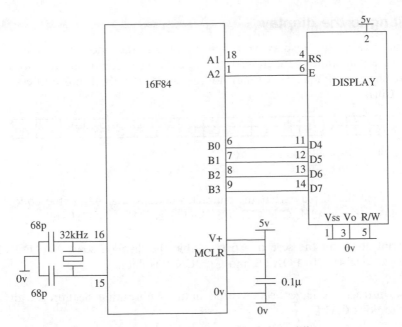

Figure 10.2 Driving the alpha numeric display with 6 control lines

Configuring the display

Before writing to the display you first of all have to configure it. That means tell it if you are:

(a) using a 4 bit or 8 bit Microcontroller,
(b) using a 1 or 2 line display,
(c) using a character font size of 5×10 or 5×7 dots,
(d) turning the display on or off,
(e) turning the cursor on or off,
(f) incrementing the cursor or not. The cursor position increments after a character has been written to the display.

In the program shown below the display has been set up in the Configuration Section with Function Set at 32H to use a 4 bit Microcontroller with a 2 line display and Font size of 5×7 dots. The Display is turned on and Cursor turned off with 0CH and the Cursor set to increment with 06H. This information was obtained from the display data sheet.

Writing to the display

- To write to the display you first of all set the address of the cursor (where you want the character to appear). The Cursor address locations are shown in Figure 10.3 Line1 address starts at 80H. Line2 address starts at C0H.

80	81	82	83	84	85	86	87	88	89	8A	8B	8C	8D	8E	8F
C0	C1	C2	C3	C4	C5	C6	C7	C8	C9	CA	CB	CC	CD	CE	CF

Figure 10.3 Cursor address location

- Then tell the display what the character code is, e.g. A has the code 41H, B has the code 42H, C is 43H, 0 is 30H, 1 is 31H, 2 is 32H etc.

To print an A on the screen – first enable the display, send 2 to PORTA, send the code 41H to PORTB and CLOCK this data.

These instructions have been written in the Subroutine Section so all you have to do is CALL A.

To write HELLO on the display the program would be:

CALL H
CALL E
CALL L
CALL L
CALL O

Program example

The program below is the listing to spell out MICROCONTROLLERS AT THE MMU.

Then CONTACT DAVE SMITH. Together with the time delays.

;ANHEAD84.ASM Header for the alpha numeric display using 6 I/O

TMR0	EQU	1	;means TMR0 is file 1.
STATUS	EQU	3	;means STATUS is file 3.
PORTA	EQU	5	;means PORTA is file 5.
PORTB	EQU	6	;means PORTB is file 6.
TRISA	EQU	85H	;TRISA (the PORTA I/O selection) is file 85H
TRISB	EQU	86H	;TRISB (the PORTB I/O selection) is file 86H
OPTION_R	EQU	81H	;the OPTION register is file 81H

```
ZEROBIT      EQU      2          ;means ZEROBIT is bit 2.
COUNT        EQU      0CH        ;COUNT is file 0C, a register to count events.
;**********************************************************
```

```
LIST         P=16F84             ;we are using the 16F84.
ORG          0                   ;the start address in memory is 0
GOTO         START               ;goto start!
```

```
;**************************************************************************
; Configuration Bits
```

```
__CONFIG H'3FF0'    ;selects LP oscillator, WDT off, PUT on,
                    ;Code Protection disabled.
```

```
;******************************************************
; SUBROUTINE SECTION.
```

```
;3 SECOND DELAY
DELAY3       CLRF     TMR0                     ;Start TMR0
LOOPA        MOVF     TMR0,W                   ;Read TMR0 into W
             SUBLW    .96                      ;TIME - W
             BTFSS    STATUS,ZEROBIT           ;Check TIME-W = 0
             GOTO     LOOPA
             RETLW    0                        ;return after TMR0 = 96
```

```
;P1 SECOND DELAY
DELAYP1      CLRF     TMR0                     ;Start TMR0
LOOPC        MOVF     TMR0,W                   ;Read TMR0 into W
             SUBLW    .3                       ;TIME - W
             BTFSS    STATUS,ZEROBIT           ;Check TIME-W = 0
             GOTO     LOOPC
             RETLW    0                        ;return after TMR0 = 3
```

```
CLOCK        BSF      PORTA,2
             NOP
             BCF      PORTA,2
             NOP
             RETLW    0
;***********************************************************
A            MOVLW    2                        ;enables the display
             MOVWF    PORTA
             MOVLW    4H
```

```
            MOVWF       PORTB
            CALL        CLOCK
            MOVLW       1H              ;41 is code for A
            MOVWF       PORTB
            CALL        CLOCK           ;clock character onto display.
            RETLW       0

BB          MOVLW       2               ;enables the display
            MOVWF       PORTA
            MOVLW       4H
            MOVWF       PORTB
            CALL        CLOCK
            MOVLW       2H              ;42 is code for B
            MOVWF       PORTB
            CALL        CLOCK           ;clock character onto display.
            RETLW       0

C           MOVLW       2               ;enables the display
            MOVWF       PORTA
            MOVLW       4H
            MOVWF       PORTB
            CALL        CLOCK
            MOVLW       3H
            MOVWF       PORTB
            CALL        CLOCK           ;clock character onto display.
            RETLW       0

D           MOVLW       2               ;enables the display
            MOVWF       PORTA
            MOVLW       4H
            MOVWF       PORTB
            CALL        CLOCK
            MOVLW       4H
            MOVWF       PORTB
            CALL        CLOCK           ;clock character onto display.
            RETLW       0

E           MOVLW       2               ;enables the display
            MOVWF       PORTA
            MOVLW       4H
            MOVWF       PORTB
            CALL        CLOCK
            MOVLW       5H
```

```
              MOVWF      PORTB
              CALL       CLOCK        ;clock character onto display.
              RETLW      0

F             MOVLW      2            ;enables the display
              MOVWF      PORTA
              MOVLW      4H
              MOVWF      PORTB
              CALL       CLOCK
              MOVLW      6H
              MOVWF      PORTB
              CALL       CLOCK        ;clock character onto display.
              RETLW      0

G             MOVLW      2            ;enables the display
              MOVWF      PORTA
              MOVLW      4H
              MOVWF      PORTB
              CALL       CLOCK
              MOVLW      7H
              MOVWF      PORTB
              CALL       CLOCK        ;clock character onto display.
              RETLW      0

H             MOVLW      2            ;enables the display
              MOVWF      PORTA
              MOVLW      4H
              MOVWF      PORTB
              CALL       CLOCK
              MOVLW      8H
              MOVWF      PORTB
              CALL       CLOCK        ;clock character onto display.
              RETLW      0

I             MOVLW      2            ;enables the display
              MOVWF      PORTA
              MOVLW      4H
              MOVWF      PORTB
              CALL       CLOCK
              MOVLW      9H
              MOVWF      PORTB
              CALL       CLOCK        ;clock character onto display.
              RETLW      0
```

```
J       MOVLW    2            ;enables the display
        MOVWF    PORTA
        MOVLW    4H
        MOVWF    PORTB
        CALL     CLOCK
        MOVLW    0AH
        MOVWF    PORTB
        CALL     CLOCK        ;clock character onto display.
        RETLW    0


K       MOVLW    2            ;enables the display
        MOVWF    PORTA
        MOVLW    4H
        MOVWF    PORTB
        CALL     CLOCK
        MOVLW    0BH
        MOVWF    PORTB
        CALL     CLOCK        ;clock character onto display.
        RETLW    0


L       MOVLW    2            ;enables the display
        MOVWF    PORTA
        MOVLW    4H
        MOVWF    PORTB
        CALL     CLOCK
        MOVLW    0CH
        MOVWF    PORTB
        CALL     CLOCK        ;clock character onto display.
        RETLW    0


M       MOVLW    2            ;enables the display
        MOVWF    PORTA
        MOVLW    4H
        MOVWF    PORTB
        CALL     CLOCK
        MOVLW    0DH
        MOVWF    PORTB
        CALL     CLOCK        ;clock character onto display.
        RETLW    0


N       MOVLW    2            ;enables the display
        MOVWF    PORTA
        MOVLW    4H
```

```
            MOVWF      PORTB
            CALL       CLOCK          ;clock character onto display.
            MOVLW      0EH
            MOVWF      PORTB
            CALL       CLOCK          ;clock character onto display.
            RETLW      0


O           MOVLW      2              ;enables the display
            MOVWF      PORTA
            MOVLW      4H
            MOVWF      PORTB
            CALL       CLOCK
            MOVLW      0FH
            MOVWF      PORTB
            CALL       CLOCK          ;clock character onto display.
            RETLW      0


P           MOVLW      2
            MOVWF      PORTA
            MOVLW      5H
            MOVWF      PORTB
            CALL       CLOCK
            MOVLW      0H
            MOVWF      PORTB
            CALL       CLOCK          ;clock character onto display.
            RETLW      0


Q           MOVLW      2
            MOVWF      PORTA
            MOVLW      5H
            MOVWF      PORTB
            CALL       CLOCK
            MOVLW      1H
            MOVWF      PORTB
            CALL       CLOCK          ;clock character onto display.
            RETLW      0


R           MOVLW      2
            MOVWF      PORTA
            MOVLW      5H
            MOVWF      PORTB
            CALL       CLOCK
            MOVLW      2H
```

```
        MOVWF    PORTB
        CALL     CLOCK        ;clock character onto display.
        RETLW    0

S       MOVLW    2
        MOVWF    PORTA
        MOVLW    5H
        MOVWF    PORTB
        CALL     CLOCK
        MOVLW    3H
        MOVWF    PORTB
        CALL     CLOCK        ;clock character onto display.
        RETLW    0

T       MOVLW    2
        MOVWF    PORTA
        MOVLW    5H
        MOVWF    PORTB
        CALL     CLOCK
        MOVLW    4H
        MOVWF    PORTB
        CALL     CLOCK        ;clock character onto display.
        RETLW    0

U       MOVLW    2
        MOVWF    PORTA
        MOVLW    5H
        MOVWF    PORTB
        CALL     CLOCK
        MOVLW    5H
        MOVWF    PORTB
        CALL     CLOCK        ;clock character onto display.
        RETLW    0

V       MOVLW    2
        MOVWF    PORTA
        MOVLW    5H
        MOVWF    PORTB
        CALL     CLOCK
        MOVLW    6H
        MOVWF    PORTB
        CALL     CLOCK        ;clock character onto display.
        RETLW    0
```

```
WW      MOVLW       2
        MOVWF       PORTA
        MOVLW       5H
        MOVWF       PORTB
        CALL        CLOCK
        MOVLW       7H
        MOVWF       PORTB
        CALL        CLOCK       ;clock character onto display.
        RETLW       0


X       MOVLW       2
        MOVWF       PORTA
        MOVLW       5H
        MOVWF       PORTB
        CALL        CLOCK
        MOVLW       8H
        MOVWF       PORTB
        CALL        CLOCK       ;clock character onto display.
        RETLW       0


Y       MOVLW       2
        MOVWF       PORTA
        MOVLW       5H
        MOVWF       PORTB
        CALL        CLOCK
        MOVLW       9H
        MOVWF       PORTB
        CALL        CLOCK       ;clock character onto display.
        RETLW       0


Z       MOVLW       2
        MOVWF       PORTA
        MOVLW       5H
        MOVWF       PORTB
        CALL        CLOCK
        MOVLW       0AH
        MOVWF       PORTB
        CALL        CLOCK       ;clock character onto display.
        RETLW       0


NUM0    MOVLW       2           ;enables the display
        MOVWF       PORTA
        MOVLW       3H
        MOVWF       PORTB
```

```
        CALL        CLOCK
        MOVLW       0H
        MOVWF       PORTB
        CALL        CLOCK       ;clock character onto display.
        RETLW       0

NUM1    MOVLW       2           ;enables the display
        MOVWF       PORTA
        MOVLW       3H
        MOVWF       PORTB
        CALL        CLOCK
        MOVLW       1H
        MOVWF       PORTB
        CALL        CLOCK       ;clock character onto display.
        RETLW       0

NUM2    MOVLW       2           ;enables the display
        MOVWF       PORTA
        MOVLW       3H
        MOVWF       PORTB
        CALL        CLOCK
        MOVLW       2H
        MOVWF       PORTB
        CALL        CLOCK       ;clock character onto display.
        RETLW       0

NUM3    MOVLW       2           ;enables the display
        MOVWF       PORTA
        MOVLW       3H
        MOVWF       PORTB
        CALL        CLOCK
        MOVLW       3H
        MOVWF       PORTB
        CALL        CLOCK       ;clock character onto display.
        RETLW       0

NUM4    MOVLW       2           ;enables the display
        MOVWF       PORTA
        MOVLW       3H
        MOVWF       PORTB
        CALL        CLOCK       ;clock character onto display.
        MOVLW       4H
```

```
                MOVWF     PORTB
                CALL      CLOCK        ;clock character onto display.
                RETLW     0

NUM5            MOVLW     2            ;enables the display
                MOVWF     PORTA
                MOVLW     3H
                MOVWF     PORTB
                CALL      CLOCK
                MOVLW     5H
                MOVWF     PORTB
                CALL      CLOCK        ;clock character onto display.
                RETLW     0

NUM6            MOVLW     2            ;enables the display
                MOVWF     PORTA
                MOVLW     3H
                MOVWF     PORTB
                CALL      CLOCK
                MOVLW     6H
                MOVWF     PORTB
                CALL      CLOCK        ;clock character onto display.
                RETLW     0

NUM7            MOVLW     2            ;enables the display
                MOVWF     PORTA
                MOVLW     3H
                MOVWF     PORTB
                CALL      CLOCK
                MOVLW     7H
                MOVWF     PORTB
                CALL      CLOCK        ;clock character onto display.
                RETLW     0

NUM8            MOVLW     2            ;enables the display
                MOVWF     PORTA
                MOVLW     3H
                MOVWF     PORTB
                CALL      CLOCK
                MOVLW     8H
                MOVWF     PORTB
                CALL      CLOCK        ;clock character onto display.
                RETLW     0
```

```
NUM9       MOVLW      2                ;enables the display
           MOVWF      PORTA
           MOVLW      3H
           MOVWF      PORTB
           CALL       CLOCK
           MOVLW      9H
           MOVWF      PORTB
           CALL       CLOCK            ;clock character onto display.
           RETLW      0

GAP        MOVLW      2
           MOVWF      PORTA
           MOVLW      2H
           MOVWF      PORTB
           CALL       CLOCK
           MOVLW      0H
           MOVWF      PORTB
           CALL       CLOCK            ;clock character onto display.
           RETLW      0

DOT        MOVLW      2
           MOVWF      PORTA
           MOVLW      2H
           MOVWF      PORTB
           CALL       CLOCK
           MOVLW      0EH
           MOVWF      PORTB
           CALL       CLOCK            ;clock character onto display.
           RETLW      0

CLRDISP    CLRF       PORTA
           MOVLW      0H
           MOVWF      PORTB
           CALL       CLOCK            ;clock character onto display.
           MOVLW      1
           MOVWF      PORTB
           CALL       CLOCK
           CALL       DELAYP1
           RETLW      0
```

;**
; CONFIGURATION SECTION.

```
START      BSF        STATUS,5         ;Turns to Bank1.
           MOVLW      B'00000000'      ;PORTA is O/P
           MOVWF      TRISA
```

```
        MOVLW       B'00000000'
        MOVWF       TRISB           ;PORTB is OUTPUT

        MOVLW       B'00000111'     ;Prescaler is /256
        MOVWF       OPTION_R        ;TIMER is 1/32 secs.

        BCF         STATUS,5        ;Return to Bank0.
        CLRF        PORTA           ;Clears PortA.
        CLRF        PORTB           ;Clears PortB.

;Display Configuration
        MOVLW       03H             ;FUNCTION SET
        MOVWF       PORTB           ;8bit data (default)
        CALL        CLOCK

        CALL        DELAYP1         ;wait for display

        MOVLW       02H             ;FUNCTION SET
        MOVWF       PORTB           ;change to 4bit
        CALL        CLOCK           ;clock in data

        CALL        DELAYP1         ;wait for display
        MOVLW       02H             ;FUNCTION SET
        MOVWF       PORTB           ;must repeat command
        CALL        CLOCK           ;clock in data

        CALL        DELAYP1         ;wait for display
        MOVLW       08H             ;4 bit micro
        MOVWF       PORTB           ;using 2 line display.
        CALL        CLOCK           ;clock in data

        CALL        DELAYP1
        MOVLW       0H              ;Display on, cursor off
        MOVWF       PORTB           ;0CH
        CALL        CLOCK
        MOVLW       0CH
        MOVWF       PORTB
        CALL        CLOCK

        CALL        DELAYP1
        MOVLW       0H              ;Increment cursor, 06H
        MOVWF       PORTB
        CALL        CLOCK
        MOVLW       6H
        MOVWF       PORTB
        CALL        CLOCK
```

```
;**********************************************************
;Program starts now.

BEGIN      CALL       CLRDISP
           CLRF       PORTA
           MOVLW      8H              ;Cursor at top left, 80H
           MOVWF      PORTB
           CALL       CLOCK
           MOVLW      0H
           MOVWF      PORTB
           CALL       CLOCK


           CALL       M               ;display M
           CALL       DELAYP1         ;wait 0.1 seconds
           CALL       I               ;display I
           CALL       DELAYP1         ;wait 0.1 seconds
           CALL       C               ;Etc.
           CALL       DELAYP1
           CALL       R
           CALL       DELAYP1
           CALL       O
           CALL       DELAYP1
           CALL       C
           CALL       DELAYP1
           CALL       O
           CALL       DELAYP1
           CALL       N
           CALL       DELAYP1
           CALL       T
           CALL       DELAYP1
           CALL       R
           CALL       DELAYP1
           CALL       O
           CALL       DELAYP1
           CALL       L
           CALL       DELAYP1
           CALL       L
           CALL       DELAYP1
           CALL       E
           CALL       DELAYP1
           CALL       R
           CALL       DELAYP1
           CALL       S
           CALL       DELAYP1
```

```
CLRF      PORTA
MOVLW     0CH         ;Cursor on second line, C3
MOVWF     PORTB
CALL      CLOCK
MOVLW     3H
MOVWF     PORTB
CALL      CLOCK

CALL      A
CALL      DELAYP1
CALL      T
CALL      DELAYP1
CALL      GAP
CALL      T
CALL      DELAYP1
CALL      H
CALL      DELAYP1
CALL      E
CALL      DELAYP1
CALL      GAP
CALL      M
CALL      DELAYP1
CALL      M
CALL      DELAYP1
CALL      U
CALL      DELAYP1

CALL      DELAY3      ;wait 3 seconds

CALL      CLRDISP

MOVLW     8H          ;Cursor at top left, 80H
MOVWF     PORTB
CALL      CLOCK
MOVLW     0H
MOVWF     PORTB
CALL      CLOCK

CALL      C
CALL      DELAYP1
CALL      O
CALL      DELAYP1
CALL      N
CALL      DELAYP1
CALL      T
```

```
        CALL        DELAYP1
        CALL        A
        CALL        DELAYP1
        CALL        C
        CALL        DELAYP1
        CALL        T
        CALL        DELAYP1
        CLRF        PORTA
        MOVLW       0CH              ;Cursor on 2nd line
        MOVWF       PORTB
        CALL        CLOCK
        MOVLW       3H
        MOVWF       PORTB
        CALL        CLOCK

        CALL        D
        CALL        DELAYP1
        CALL        A
        CALL        DELAYP1
        CALL        V
        CALL        DELAYP1
        CALL        E
        CALL        DELAYP1
        CALL        GAP
        CALL        DELAYP1
        CALL        S
        CALL        DELAYP1
        CALL        M
        CALL        DELAYP1
        CALL        I
        CALL        DELAYP1
        CALL        T
        CALL        DELAYP1
        CALL        H

        CALL        DELAY3           ;wait 3 seconds

        GOTO        BEGIN
END
```

Program operation

- PORTA and PORTB are configured as outputs in the CONFIGURATION
  SECTION.

Display configuration

- In the Display Configuration Section, the Register Select (R/S) line, A1on the microcontroller, is set low by CLRF PORTA in the Configuration Section.
- R/S = 0 ensures that the data to the display will change the registers. Later R/S = 1 writes the characters to the display.
- The display is expecting its data to arrive via 8 lines, but to save I/O lines we will use 4 and write them twice. The code to do this and also tell the driver chip the display is a two line display is:

```
MOVLW    03H        ;FUNCTION SET
MOVWF    PORTB      ;8bit data (default)
CALL     CLOCK

CALL     DELAYP1    ;wait for display

MOVLW    02H        ;FUNCTION SET
MOVWF    PORTB      ;change to 4bit
CALL     CLOCK      ;clock in data

CALL     DELAYP1    ;wait for display
MOVLW    02H        ;FUNCTION SET
MOVWF    PORTB      ;must repeat command
CALL     CLOCK      ;clock in data

CALL     DELAYP1    ;wait for display
MOVLW    08H        ;4 bit micro
MOVWF    PORTB      ;using 2 line display.
CALL     CLOCK      ;clock in data
```

The data is set up on PORTB using B0,1,2 and 3. As in

```
MOVLW    03H        ;FUNCTION SET
MOVWF    PORTB
```

This data is then clocked into the display by pulsing the Enable line, (E, A2 on the micro) high and then low with:

```
CLOCK    BSF      PORTA,2
         NOP
         BCF      PORTA,2
         NOP
         RETLW    0
```

CALL DELAYP1 , waits for 0.1 seconds to give the display time to activate before continuing.

When the display has been configured to: Turn on, switch the cursor off, and increment the cursor after every character write. We are then ready to write to the display.

Writing to the display

- The display is cleared if required with:

  CALL CLRDISP

- The address of the character is first written to the display, say, the 80H position (top left hand corner).

```
CLRF      PORTA
MOVLW     8H          ;Cursor at top left, 80H
MOVWF     PORTB
CALL      CLOCK
MOVLW     0H
MOVWF     PORTB
CALL      CLOCK
```

Notice the 8 is sent first followed by the 0.

To write to the position mid-way along the top line the address would be 88H. So the 80H in the code above would be replaced by 88H.

- In order to write the letter 'M' in the display at the position defined. We CALL M and use the code 4DH, NB. Send the 4 first followed by the D. The Register Select Line, R/S, A1 on the micro, is set to 1 for the character write option. The code is:

```
M         MOVLW     2          ;enables the display
          MOVWF     PORTA      ;sets A1=1
          MOVLW     4H         ;send data 4
          MOVWF     PORTB
          CALL      CLOCK
          MOVLW     0DH        ;send data D
          MOVWF     PORTB
          CALL      CLOCK      ;clock character 'M' onto display.
          RETLW     0
```

In this way any one of the 240 characters available can be shown on the display.

The program continues by printing out the rest of the message. A delay of 0.1 seconds is maintained after printing each character to give the effect of the message being typed out.

All the Capital Letters and numbers 0 to 9 have been included in the header so you can easily enter your own message.

The complete character set for the display showing all 240 characters is illustrated in Figure 10.4.

Displaying a number

Suppose we wish to display a number thrown by a dice, for example a 4. We could use the instruction CALL NUM4, but we would not have known previously that the number was going to be a 4. The throw of the dice would be stored in a user file called, say, THROW and THROW would then have 4 in it.

Now the code for 0 is 30H
The code for 1 is 31H
The code for 2 is 32H
Etc.
If we wanted to display the number 4 the code is:

```
NUM4    MOVLW    2          ;enables the display
        MOVWF    PORTA
        MOVLW    3H         ;34H is the code for 4
        MOVWF    PORTB
        CALL     CLOCK
        MOVLW    4H
        MOVWF    PORTB
        CALL     CLOCK      ;clock character onto display.
        RETLW    0
```

If the 4 is in the file THROW, we can display this with the code:

```
        MOVLW    2          ;enables the display
        MOVWF    PORTA
        MOVLW    3H
        MOVWF    PORTB
```

```
CALL      CLOCK
MOVF      THROW,W      ;number comes from the file
MOVWF     PORTB
CALL      CLOCK        ;clock character onto display.
RETLW     0
```

Notice how the value of the number now has come from the file.

This code would then display any number in the file **THROW**.

If you measured a temperature as 27°C, you would probably store the 2 in a file TEMPTENS (tens of degrees) and the 7 in a file TEMPUNIT (units of degrees).

You can then modify the code above to display:

THE TEMPERATURE

IS 27°C.

The 'I' would be located at address C5H on the display. The temperature would then be written at locations C8H and C9H. There would be no need to rewrite the message just rewrite the temperature as it changed, after first moving the cursor to address C8H.

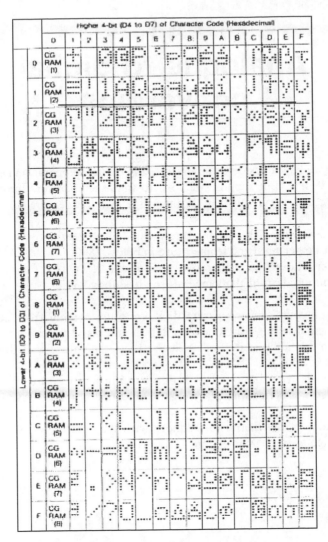

Figure 10.4 Alpha numeric display character set

11
Analogue to digital conversion

Up to now we have considered inputs as being digital in operation i.e. the input is either a 0 or 1. But suppose we wish to make temperature measurements, but not just hot or cold (1 or 0). We may for example require to:

(a) Sound a buzzer if the temperature drops below freezing.
(b) Turn a heater on if the temperature is below 18°C.
(c) Turn on a fan if the temperature goes above 25°C.
(d) Turn on an alarm if the temperature goes above 30°C.

We could of course have separate digital inputs, coming from comparator circuits for each setting. But a better solution is to use 1 input connected to an analogue to digital converter and measure the temperature with that.

Figure 11.1 shows a basic circuit for measuring temperature. It consists of a fixed resistor in series with a thermistor (a temperature sensitive resistor).

The resistance of the thermistor changes with temperature causing a change in the voltage at point X in Figure 11.1.

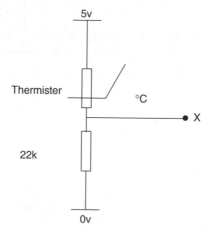

Figure 11.1 Temperature measuring circuit

As the temperature rises the voltage at X rises.
As the temperature decreases the voltage at X reduces.

We need to know the relationship between the temperature of the thermistor and the voltage at X. A simple way of doing this would be to place the thermistor in a cup of boiling water (100°C) and measure the voltage at X. As the water cools corresponding readings of temperature and voltage can be taken. If needed a graph of these temperature and voltage readings could be plotted.

Making an A/D reading

In the initial example let us suppose:

- 0°C gave a voltage reading of 0.6v
- 18°C gave a reading of 1.4v
- 25°C gave a reading of 2.4v
- 30°C gave a reading of 3.6v

The microcontroller would read these voltages and convert them to an 8-bit number where 0v is 0 and 5v is 255. I.e. a reading of 51 per volt or a resolution of 1/51v, i.e. 1 bit is 19.6mv.

So 0°C = 0.6v = reading of 31 (0.6 × 51 = 30.6)
18°C = 1.4v = 71 (1.4 × 51 = 71.4)
25°C = 2.4v = 122 (2.4 × 51 = 122.4)
30°C = 3.6v = 184 (3.6 × 51 = 183.6)

If we want to know when the temperature is above 30°C the microcontroller looks to see if the A/D reading is above 184. If it is, switch on the alarm, if not keep the alarm off. In a similar way any other temperature can be investigated – not just the ones listed. With our 8 bits we have 255 different temperatures we can choose from. The PIC 16C773 and PIC 16C774 have 12-bit A/D converters and can have 4096 different temperature points.

Analogue to Digital conversion was introduced to the PIC Microcontrollers with the family called 16C7X devices: 16C71, 16C73 and 16C74. Table 11.1 shows some of the specifications of these devices.

Table 11.1 16C7X Device specifications

Device	I/O	A/D Channels	Program Memory	Data Memory	Current Source/Sink
16C71	13	4	1k	36	25mA
16C73	22	5	4k	192	25mA
16C74	33	8	4k	192	25mA

This family of devices has now been superceded by the 16F87X devices shown in Table 11.2.

Table 11.2 16F87X Devices

Device	I/O	A/D Channels	Program Memory	Data Memory	Current Source/Sink
16F870	22	5	2k	128	25mA
16F871	33	8	2k	128	25mA
16F872	22	5	2k	128	25mA
16F873	22	5	4k	192	25mA
16F874	33	8	4k	192	25mA
16F876	22	5	8k	368	25mA
16F877	33	5	8k	368	25mA

The device I shall consider in this section is the 16F818. The Device Family Specifications are shown in Table 11.3.

Table 11.3 16F818/9 Device specifications

Device	I/O	A/D Channels	Program Memory	Data Memory	Current Source/Sink
16F818	16	5	1k	128	25mA
16F819	16	5	2k	256	25mA

The 16F818 device needs extra registers that the 16F84 does not have, to handle the A/D processing.

The 16F818 has 5 Analogue Inputs AN0, AN1, AN2, AN3 and AN4.

Configuring the A/D device

In order to make an analogue measurement we have to configure the device. HEAD818.ASM has to have the CONFIGURATION SECTION changed to make some of the PORTA inputs Analogue inputs. PORTB has been set as an output port.

To configure the 16F818 for A–D measurements three registers need to be set up.

- ADCON0
- ADCON1
- ADRES

ADCON0

The first of the A/D registers, ADCON0 is A to D Control Register 0.

ADCON0 is used to:

- Switch the A/D converter on with ADON, bit0. This bit turns the A/D on when set and off when clear. The A/D once it is turned on can be left on all of the time but it does draw a current of 90μA, compared to the rest of the microcontroller which draws a current of 15μA.
- Instruct the microcontroller to execute a conversion by setting the GO/DONE bit, bit2. When the GO/DONE bit is set the micro does an A/D conversion. When the conversion is complete the hardware clears the GO/DONE bit. This bit can be read to determine when the result is ready.
- Set the particular channel (input) to make the measurement from. This is done with two Channel Select bits, CHS0, CHS1 and CHS2, bits 3, 4 and 5.

The Register ADCON0 is shown in Figure 11.2.

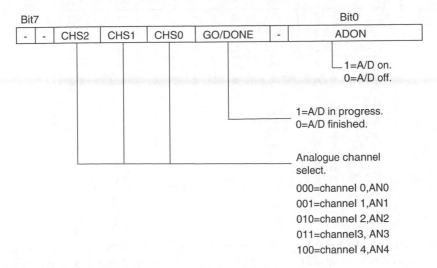

Figure 11.2 ADCON0 Register

ADCON1

In ADCON1, A to D Conversion Register 1, only bits 0, 1, 2 and 3 are used.

They are the Port Configuration bits, PCFG0, PCFG1, PCFG2, and PCFG3 that determine which of the pins on PORTA will be analogue inputs and which will be digital.

The ADCON1 register is illustrated in Figure 11.3 and the corresponding Analogue and Digital inputs are shown in Table 11.4.

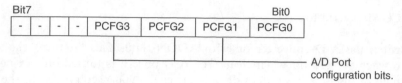

Figure 11.3 ADCON1 Register

Table **11.4** ADCON1 Port configuration

PCFG	AN4	AN3	AN2	AN1	AN0	Vref+	Vref−
0000	A	A	A	A	A	Vdd	Vss
0001	A	Vref+	A	A	A	AN3	Vss
0010	A	A	A	A	A	Vdd	Vss
0011	A	Vref+	A	A	A	AN3	Vss
0100	D	A	D	A	A	Vdd	Vss
0101	D	Vref+	D	A	A	AN3	Vss
011X	D	D	D	D	D	Vdd	Vss
1000	A	Vref+	Vref−	A	A	AN3	AN2
1001	A	A	A	A	A	Vdd	Vss
1010	A	Vref+	A	A	A	AN3	Vss
1011	A	Vref+	Vref−	A	A	AN3	AN2
1100	A	Vref+	Vref−	A	A	AN3	AN2
1101	D	Vref+	Vref−	A	A	AN3	AN2
1110	D	D	D	D	A	Vdd	Vss
1111	D	Vref+	Vref−	D	A	AN3	AN2

As mentioned previously the microcontroller will convert an analogue voltage between 0 and 5v to a digital number between 0 and 255. But suppose our analogue readings of say, temperature, go from 0.6v representing a temperature of 0°C to 3.6v representing a temperature of 30°C. It would make sense to have our analogue range go from 0.6v to 3.6v. We can set this by using two reference voltages. One at the low setting of 0.6v called Vref−, connected to AN2. The other setting of 3.6v for Vref+, connected to AN3. The two right hand columns in Table 11.4 show that PCFG Set at 1000 will set the A/D configuration using AN3 and AN2 as the reference voltages. In this book I have not used any reference voltages but have used 5v, Vdd and 0v. Vss as the references.

ADRES

• The third register is ADRES, the A to D RESult register. This is the file where the result of the A/D conversion is stored. If several measurements require storing then the number in ADRES needs to be transferred to a user file before it is overwritten with the next measurement. The 16F818 micro is a 10 bit A/D. The top 8 bits are stored in ADRESH and the lower 2 bits in ADRESL. In this book I am only using 8 bits and have called the file ADRES.

Analogue header for the 16F818

```
;HEAD818A.ASM for 16F818.        This sets PORTA as analogue/digital
;                                INPUTs.
;                                PORTB is an OUTPUT.
;                                Internal oscillator of 31.25kHz chosen
;                                The OPTION register is set to /256 giving
;                                timing pulses 32.768ms.
;                                1second and 0.5 second delays are
;                                included in the subroutine section.

;************************************************************

; EQUATES SECTION

TMR0        EQU    1            ;means TMR0 is file 1.
STATUS      EQU    3            ;means STATUS is file 3.
PORTA       EQU    5            ;means PORTA is file 5.
PORTB       EQU    6            ;means PORTB is file 6.
ZEROBIT     EQU    2            ;means ZEROBIT is bit 2.
ADCON0      EQU    1FH          ;A/D Configuration reg.0
ADCON1      EQU    9FH          ;A/D Configuration reg.1
ADRES       EQU    1EH          ;A/D Result register.
CARRY       EQU    0            ;CARRY IS BIT 0.
TRISA       EQU    85H          ;PORTA Configuration Register
TRISB       EQU    86H          ;PORTB Configuration Register
OPTION_R    EQU    81H          ;Option Register
OSCCON      EQU    8FH          ;Oscillator control register.
COUNT       EQU    20H          ;COUNT a register to count events.

;************************************************************

        LIST        P=16F818     ;we are using the 16F818.
        ORG         0            ;the start address in memory is 0
        GOTO        START        ;goto start!
```

```
;*******************************************************
;Configuration Bits

__CONFIG H'3F10'      ;sets INTRC-A6 is port I/O, WDT off, PUT
                      ;on, MCLR tied to VDD A5 is I/O
                      ;BOD off, LVP disabled, EE protect disabled,
                      ;Flash Program Write disabled,
                      ;Background Debugger Mode disabled, CCP
                      ;function on B2,
                      ;Code Protection disabled.


;*******************************************************

;SUBROUTINE SECTION.

;0.1 second delay, actually 0.099968s
DELAYP1 CLRF     TMR0                  ;START TMR0.
LOOPB   MOVF     TMR0,W                ;READ TMR0 INTO W.
        SUBLW    .3                    ;TIME-3
        BTFSS    STATUS,ZEROBIT        ;Check TIME-W = 0
        GOTO     LOOPB                 ;Time is not = 3.
        NOP                            ;add extra delay
        NOP
        RETLW    0                     ;Time is 3, return.

;0.5 second delay.
DELAYP5    MOVLW    .5
           MOVWF    COUNT


LOOPC      CALL     DELAYP1
           DECFSZ   COUNT
           GOTO     LOOPC
           RETLW    0


;1 second delay.
DELAY1     MOVLW    .10
           MOVWF    COUNT


LOOPA      CALL     DELAYP1
           DECFSZ   COUNT
           GOTO     LOOPA
           RETLW    0
```

```
;**********************************************************
;CONFIGURATION SECTION.

START      BSF            STATUS,5        ;Turns to Bank1.

           MOVLW          B11111111'      ;8 bits of PORTA are I/P
           MOVWF          TRISA

           MOVLW          B'00000100'     ;A0,A1 and A3 are analogue.
           MOVWF          ADCON1

           MOVLW          B'00000000'
           MOVWF          TRISB           ;PORTB is OUTPUT

           MOVLW          B'00000000'
           MOVWF          OSCCON          ;oscillator 31.25kHz

           MOVLW          B'00000111'     ;Prescaler is /256
           MOVWF          OPTION_R        ;TIMER is 1/32 secs.

           BCF            STATUS,5        ;Return to Bank0.

           BSF            ADCON0,0        ;Turn ON A/D

           CLRF           PORTA           ;Clears PortA.
           CLRF           PORTB           ;Clears PortB.

;**********************************************************
;Program starts now.

END
```

Head818A.ASM explained

HEAD818A.ASM is similar in operation to HEAD818.ASM outlined in Chapter 6, with the following extras:

- The Carry Bit in the status register, that indicates if a calculation is +ve or −ve, it is bit 0 and has been equated to 0.

- In the Configuration Section A0, A1 and A3 are set as Analogue inputs, A2, A4, A5, A6 and A7 are set up as digital inputs with:

MOVLW   B'00000100'
MOVWF   ADCON1

The A/D converter is switched on with:

BSF   ADCON0,0

A/D Conversion – example, a temperature sensitive switch

To introduce the working of the A/D converter we will consider a simple example. i.e. Turn an LED on when the Temperature is above 25°C and turn the LED off when it is below 25°C.

The diagram for this Temperature Switch Circuit is shown in Figure 11.4.

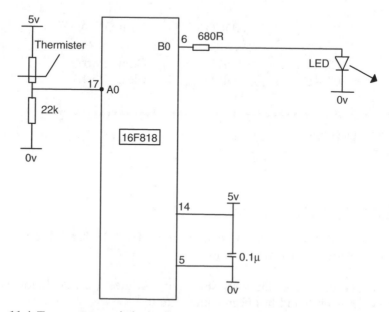

Figure 11.4 Temperature switch circuit

Taking the A/D reading

The A/D converter has been switched on in the header and it automatically looks at Channel 0 unless told otherwise. In order to make the measurement the GO/DONE bit, bit2 is set and we wait until it is cleared with:

```
        BSF       ADCON0,2   ;Take measurement, set GO/DONE
WAIT    BTFSC     ADCON0,2   ;Wait until GO/DONE is clear
        GOTO      WAIT
```

The measurement will then be in the A/D Result register, ADRES.

Determining if the temperature is above or below 25°C

Suppose the voltage on the analogue input, Channel 0, A0 is 2.4v when the temperature is 25°C. The required A/D reading for 2.4v is $2.4 \times 51 = 122$. We therefore need to know when the A/D reading is above and below 122, i.e. above and below 25°C.

Previously we have seen how to tell if a value is equal to another by subtracting and looking at the zerobit in the status register (Chapter 5).

There is another bit, bit 0 in the status register called the Carry Bit, which indicates if the result of a subtraction is +ve or −ve. If the Carry Bit is set the result was +ve, if the bit is clear the result was −ve. So we can tell if the number is above or below a defined value.

The code for this is:

```
MOVF    ADRES,W         ;Move Analogue result into W
SUBLW   .122            ;Do 122 – ADRES, i.e. 122-W
BTFSC   Status,Carry    ;Check the carry bit. Clear if ADRES>122 i.e. −ve
GOTO    TURNOFF         ;Routine to turn off LED
GOTO    TURNON          ;Routine to turn on LED
```

The analogue measurement is moved from ADRES into W where we can subtract it from 122. NB. The subtraction always does, Value − W.

The carry bit tells us if the A/D result is above or below 122.

N.B. If the result of the subtraction is zero the carry is also 1. It must be 1 or 0. Being +ve or zero does not matter in this example.

We have then found out if the result is equal to or above 122, or if it is less than 122.

When the measurement is made we then goto one of two subroutines, TURNON or TURNOFF. These subroutines are not very grand but they could easily be more complicated, even hundreds of lines long.

Program code

The full code for this Temperature Sensitive Switch Program is shown below as TEMPSENS.ASM

```
;TEMPSENS.ASM.          This sets PORTA as analogue/digital INPUTs.
;                       PORTB is an OUTPUT.
;                       Internal oscillator of 31.25kHz chosen
;                       The OPTION register is set to /256 giving timing
;                       pulses 32.768ms.
;                       1second and 0.5 second delays are included in the
;                       subroutine section.


;*********************************************************
;EQUATES SECTION


TMR0        EQU     1       ;means TMR0 is file 1.
STATUS      EQU     3       ;means STATUS is file 3.
PORTA       EQU     5       ;means PORTA is file 5.
PORTB       EQU     6       ;means PORTB is file 6.
ZEROBIT     EQU     2       ;means ZEROBIT is bit 2.
ADCON0      EQU     1FH     ;A/D Configuration reg.0
ADCON1      EQU     9FH     ;A/D Configuration reg.1
ADRES       EQU     1EH     ;A/D Result register.
CARRY       EQU     0       ;CARRY IS BIT 0.
TRISA       EQU     85H     ;PORTA Configuration Register
TRISB       EQU     86H     ;PORTB Configuration Register
OPTION_R    EQU     81H     ;Option Register
OSCCON      EQU     8FH     ;Oscillator control register.
COUNT       EQU     20H     ;COUNT a register to count events


;*********************************************************


        LIST        P=16F818    ;we are using the 16F818.
        ORG         0           ;the start address in memory is 0
        GOTO        START       ;goto start!
```

```
;********************************************************
; Configuration Bits

__CONFIG H'3F10'    ;sets INTRC-A6 is port I/O, WDT off, PUT
                    ;on, MCLR tied to VDD A5 is I/O
                    ;BOD off, LVP disabled, EE protect disabled,
                    ;Flash Program Write disabled,
                    ;Background Debugger Mode disabled, CCP
                    ;function on B2,
                    ;Code Protection disabled.

;********************************************************
;SUBROUTINE SECTION.

TURNON      BSF         PORTB,0     ;Turn on LED on B0
            GOTO        BEGIN       ;Return to monitor

TURNOFF     BCF         PORTB,0     ;Turn off LED on B0
            GOTO        BEGIN       ;Return to monitor

;********************************************************
;CONFIGURATION SECTION.

START       BSF         STATUS,5    ;Turns to Bank1.

            MOVLW       B'11111111' ;8 bits of PORTA are I/P
            MOVWF       TRISA

            MOVLW       B'00000100' ;A0,A1 and A3 are analogue.
            MOVWF       ADCON1

            MOVLW       B'00000000'
            MOVWF       TRISB       ;PORTB is OUTPUT

            MOVLW       B'00000000'
            MOVWF       OSCCON      ;oscillator 31.25kHz

            MOVLW       B'00000111' ;Prescaler is /256
            MOVWF       OPTION_R    ;TIMER is 1/32 secs.

            BCF         STATUS,5    ;Return to Bank0.

            BSF         ADCON0,0    ;Turn ON A/D

            CLRF        PORTA       ;Clears PortA.
            CLRF        PORTB       ;Clears PortB.
```

```
;************************************************************
;Program starts now.
BEGIN    BSF        ADCON0,2   ;Take measurement, set GO/DONE
WAIT     BTFSC      ADCON0,2   ;Wait until GO/DONE is clear
         GOTO       WAIT

         MOVF       ADRES,W    ;Move Analogue result into W
         SUBLW      .122       ;Do 122–ADRES, i.e. 122–W
         BTFSC      STATUS,
                    CARRY      ; Clear if ADRES > 122
         GOTO       TURNOFF    ;Routine to turn off LED
         GOTO       TURNON     ;Routine to turn on LED
END
```

Another example – a voltage indicator

Previously we have looked at a single input level. But with our 8 bit micro we could look at 255 different input levels.

Suppose we wish to use the LEDs connected to PORTB to indicate the voltage on the Analogue Input AN0. So that as the voltage increases then the number of LEDs lit also increases.

In HEAD818.ASM we have configured the micro so that the voltage reference is Vdd i.e. the 5v supply. This was done with the instructions:

```
MOVLW     B'00000100'
MOVWF     ADCON1
```

This means that 5v will give a digital reading of 255 in our 8 bit register ADRES. The resolution of this register is $5v/255 = 19.6mV$.

Suppose our LED ladder was to increment in 0.5v steps as indicated below:

Vin = 0–0.5v	All LEDs off,	$0.5v = 0.5/5 \times 255 = 25.5 = 26$
Vin = 0.5–1.0v	B0 on,	$1.0v = 1/5 \times 255 = 51$
Vin = 1.0–1.5v	B1 on,	$1.5v = 1.5/5 \times 255 = 76.5 = 77$
Vin = 1.5–2.0v	B2 on,	$2.0v = 2/5 \times 255 = 102$

Vin = 2.0–2.5v	B3 on,	$2.5v = 2.5/5 \times 255 = 127.5 = 128$
Vin = 2.5–3.0v	B4 on,	$3.0v = 3/5 \times 255 = 153$
Vin = 3.0–3.5v	B5 on,	$3.5v = 3.5/5 \times 255 = 178.5 = 179$
Vin = 3.5–4.0v	B6 on,	$4.0v = 4/5 \times 255 = 204$
Vin = 4.0–5.0v	B7 on.	

The circuit diagram for this voltage indicator is shown in Figure 11.5 and the Flowchart is shown in Figure 11.6.

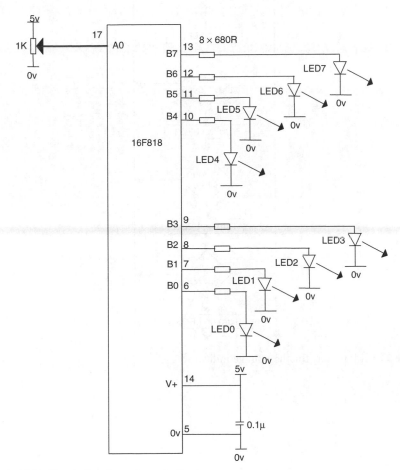

Figure 11.5 Circuit for the voltage indicator

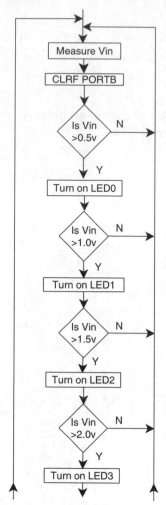

Figure 11.6 Flowchart for the voltage indicator

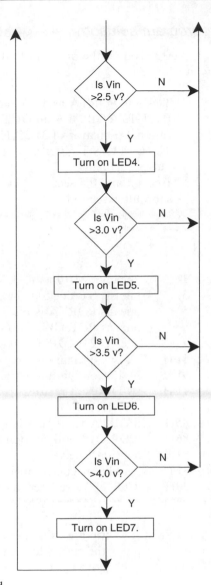

Figure 11.6 Continued

Voltage indicator, program solution

HEAD818A.ASM is altered to produce the program VOLTIND.ASM for the Voltage Indicator Circuit.

;VOLTIND.ASM.		This sets PORTA as analogue/digital
;		INPUTs. PORTB is an OUTPUT.
;		Internal oscillator of 31.25kHz chosen
;		The OPTION register is set to /256 giving timing pulses 32.768ms.
;		1second and 0.5 second delays are included in the subroutine section.

```
;********************************************************
; EQUATES SECTION

TMR0        EQU   1      ;means TMR0 is file 1.
STATUS      EQU   3      ;means STATUS is file 3.
PORTA       EQU   5      ;means PORTA is file 5.
PORTB       EQU   6      ;means PORTB is file 6.
ZEROBIT     EQU   2      ;means ZEROBIT is bit 2.
ADCON0      EQU   1FH    ;A/D Configuration reg.0
ADCON1      EQU   9FH    ;A/D Configuration reg.1
ADRES       EQU   1EH    ;A/D Result register.
CARRY       EQU   0      ;CARRY IS BIT 0.
TRISA       EQU   85H    ;PORTA Configuration Register
TRISB       EQU   86H    ;PORTB Configuration Register
OPTION_R    EQU   81H    ;Option Register
OSCCON      EQU   8FH    ;Oscillator control register.
COUNT       EQU   20H    ;COUNT a register to count events
;********************************************************

        LIST      P=16F818     ;we are using the 16F818.
        ORG       0            ;the start address in memory is 0
        GOTO      START        ;goto start!

;********************************************************
; Configuration Bits

__CONFIG H'3F10'        ;sets INTRC-A6 is port I/O, WDT off, PUT on,
                        ;MCLR tied to VDD A5 is I/O
                        ;BOD off, LVP disabled, EE protect disabled,
                        ;Flash Program Write disabled,
```

```
                              ;Background Debugger Mode disabled, CCP
                              ;function on B2,
                              ;Code Protection disabled.

;********************************************************
;CONFIGURATION SECTION.

START    BSF           STATUS,5        ;Turns to Bank1.
         MOVLW         B'00011111'     ;5bits of PORTA are I/P
         MOVWF         TRISA
         MOVLW         B'00000010'     ;A0, A1 are analogue
         MOVWF         ADCON1          ;A2, A3 are digital I/P.
         MOVLW         B'00000000'
         MOVWF         TRISB           ;PORTB is OUTPUT
         BCF           STATUS,5        ;Return to Bank0.
         MOVLW         B'00000001'     ;Turns on A/D converter,
         MOVWF         ADCON0          ;and selects channel AN0
         CLRF          PORTA           ;Clears PortA.
         CLRF          PORTB           ;Clears PortB.

;********************************************************
;Program starts now.

BEGIN    BSF           ADCON0,2        ;Take Measurement.
WAIT     BTFSC         ADCON0,2        ;Wait until reading done.
         GOTO          WAIT
         MOVF          ADRES,W         ;Move A/D Result into W
         CLRF          PORTB           ;Clear PortB.

         SUBLW         .26             ;26-,W. W is altered
         BTFSC         STATUS,CARRY    ;Is W> or <26
         GOTO          BEGIN           ;W is <26 (0.5v)

         MOVF          ADRES,W         ;Move A/D Result into W
         BSF           PORTB,0         ;Turn on B0.
         SUBLW         .51             ;51-,W. W is altered
         BTFSC         STATUS,CARRY    ;Is W> or <51
         GOTO          BEGIN           ;W is <51 (1.0v)

         MOVF          ADRES,W         ;Move A/D Result into W
         BSF           PORTB,1         ;Turn on B1.
         SUBLW         .77             ;77-,W. W is altered
         BTFSC         STATUS,CARRY    ;Is W> or <77
         GOTO          BEGIN           ;W is <77 (1.5v)
```

```
          MOVF        ADRES,W          ;Move A/D Result into W
          BSF         PORTB,2          ;Turn on B2.
          SUBLW       .102             ;102-,W. W is altered
          BTFSC       STATUS,CARRY     ;Is W> or <102
          GOTO        BEGIN            ;W is <102 (2.0v)

          MOVF        ADRES,W          ;Move A/D Result into W
          BSF         PORTB,3          ;Turn on B3.
          SUBLW       .128             ;128-,W. W is altered
          BTFSC       STATUS,CARRY     ;Is W> or <128
          GOTO        BEGIN            ;W is <128 (2.5v)

          MOVF        ADRES,W          ;Move A/D Result into W
          BSF         PORTB,4          ;Turn on B4.
          SUBLW       .153             ;153-,W. W is altered
          BTFSC       STATUS,CARRY     ;Is W> or <153
          GOTO        BEGIN            ;W is <153 (3.0v)

          MOVF        ADRES,W          ;Move A/D Result into W
          BSF         PORTB,5          ;Turn on B5.
          SUBLW       .179             ;179-,W. W is altered
          BTFSC       STATUS,CARRY     ;Is W> or <179
          GOTO        BEGIN            ;W is <179 (3.5v)

          MOVF        ADRES,W          ;Move A/D Result into W
          BSF         PORTB,6          ;Turn on B6.
          SUBLW       .204             ;204-,W. W is altered
          BTFSC       STATUS,CARRY     ;Is W> or <204
          GOTO        BEGIN            ;W is <204 (4.0v)

          BSF         PORTB,7          ;Turn on B7.
          GOTO        BEGIN
END
```

Operation of the voltage indicator program

The code to make the analogue measurement is the same as in the Temperature Switch Circuit. Once the measurement has been taken the program checks to see if the digital value of the input is >26 if it is B0 LED is switched on. The program then checks to see if the measurement is >51, if so then B1 LED is lit. If the reading is >77 then B2 LED is lit etc. When the value is less than the one being checked then the program branches back to the beginning, makes another measurement and the cycle repeats.

NB. After the A/D reading the LEDs are cleared before being turned on, in case the voltage has dropped.

To check if a reading (or any number) is > say 26.
Put the number into W.
Take W from 26 i.e. 26-W by SUBLW .26

If the result is +ve, the number is <26 and the carry bit is set in the Status Register. If the number is >26 the result is −ve and the carry bit is clear.

Problem

To check your understanding of the previous section, try this.

Turn a red LED on only when the input voltage is above 3v and turn a yellow LED on only when the input voltage is below 1v and turn a green LED on only when the voltage is between 1v and 3v.

Hint

Check for voltage >3v if true GOTO RED
If not check for voltage <1v if true GOTO YELLOW
If false then GOTO GREEN.

12
Radio transmitters and receivers

Radio circuits used to frighten me but now with the introduction of low cost modules the radio novice like myself can transmit data easily.

This section details the use of the 418 MHz Radio Transmitter and Receiver Modules (RT1-418 and RR3-418). They do not need a license to operate and there are many varieties available. The transmitters only have 3 connections, 2 power supply and one data input, the transmitting aerial is incorporated on the unit. The receiver has 4 connections, 2 power supply, 1 aerial input and 1 data output. The receiving aerial only needs to be a piece of wire about 25cm long.

The basic circuit diagram of the radio system is shown in Figure 12.1.

The microcontroller generates the data and then passes the data pulses to the transmitter. The receiver receives the data pulses and a microcontroller decodes the information and processes it.

A microcontroller-radio system could measure the temperature outside and transmit this temperature to be displayed on a unit inside.

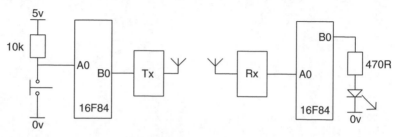

Figure 12.1 Radio data transmission system

How does it work?

The transmitter

Data is generated by the microcontroller say by pressing a switch or from a temperature sensor via the 16F818 doing an A/D conversion. Suppose this data is 27H, this would then be stored in a user file, called, say, NUMA.

So file NUMA would appear as shown in Figure 12.2.

NUMA,7	NUMA,6	NUMA,5	NUMA,4	NUMA,3	NUMA,2	NUMA,1	NUMA,0
0	0	1	0	0	1	1	1

Figure 12.2 File NUMA containing 27H

The data then needs to be passed from the micro to the data input of the transmitter. The transmitter output will then be turned on and off by the data pulses. The length of time the transmitter is on will indicate if the data was a 1, a 0 or the transmission start pulse.

I have decided to use a start bit that is 7.5ms wide, a 5ms pulse to represent a logic 1 and a 2.5ms pulse to represent a logic 0. All pulses are separated by a space of 2.5ms. The pulse train for NUMA is then as shown in Figure 12.3.

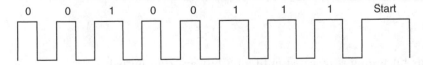

Figure 12.3 NUMA pulse train

In order to generate this train the software turns the output on for the 7.5ms start pulse, off for 2.5ms, on for 5ms for the first 1, off for 2.5ms, on for 5ms for the next logic 1, off for 2.5ms, on for 5ms for the next logic 1, off for 2.5ms, on for 2.5ms for the logic 0, etc.

To generate the data each bit in the file NUMA is tested in turn. If the bit is 0 then the output is turned on for 2.5ms, if the bit is 1 then the output is turned on for 5ms. The code for this data would be:

```
BSF      PORTB,0    ;Transmit start pulse
CALL     DELAY3     ;7.5ms Start pulse
BCF      PORTB,0    ;Transmit space
CALL     DELAY1     :Delay 2.5ms
```

```
TESTA0      BTFSC     NUMA,0      ;Test NUMA,0
            GOTO      SETA0       ;If NUMA0 = 1
            GOTO      CLRA0       ;If NUMA0 = 0

SETA0       BSF       PORTB,0     ;Transmit 1
            CALL      DELAY2      ;Delay 5ms
            GOTO      TESTA1

CLRA0       BSF       PORTB,0     ;Transmit 0
            CALL      DELAY1      ;Delay 2.5ms
            GOTO      TESTA1

TEASTA1     BCF       PORTB,0     ;Transmit space
            CALL      DELAY1
            BTFSC     NUMA,1      ;Test NUMA,1
            GOTO      SETA1       ;If NUMA0 = 1
            GOTO      CLRA1       ;If NUMA0 = 0

SETA1       BSF       PORTB,0
            CALL      DELAY2
            GOTO      TESTA2

CLRA1       BSF       PORTB,0
            CALL      DELAY1
            GOTO      TESTA2
```

●
●
●

This bit testing is repeated until all 8 bits are transmitted.

The receiver

The receiver works the opposite way round. The data is received and stored in a file NUMA. Several data bytes could be transmitted depending on how many switches are used. Or the data may be continually varying from a temperature sensor. In this example we are only looking for one byte i.e. the number 27H which was transmitted. The data is passed from the receiver to the input A0 of the microcontroller.

We wait to receive the 7.5ms start bit. When this is detected we then measure the next 8 pulses.

If a pulse is 5ms wide then a one has been transmitted and we SET the relative bit in the file NUMA. If the pulse is only 2.5ms long then we leave the bit CLEAR.

Measuring the received pulse width

Measuring the width of a pulse is a little more difficult than setting a pulse width. Consider the pulse in Figure 12.4.

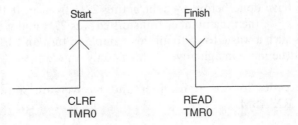

Figure 12.4 Measuring the width of a pulse

The input is continually tested until it goes high and then the timer, TMR0, is cleared to start timing. The input is continually tested until it goes low and then the value of TMR0 is read. This is done by:

MOVF   TMR0,W which puts the value of TMR0 into W.

We can then check to see if the pulse is 5ms long i.e. a logic 1, if not then a shorter pulse means a logic 0 was transmitted. If the pulse is greater than 3.5ms then it must be a logic1, at 5ms. If the pulse is less than 3.5ms then it must be a logic0. TMR0 will hold a value of 3 after a time of 3.5ms, so we check to see if the width of the pulse is greater or less than 3.

The code for this is:

```
TESTA0H    BTFSS      PORTA,0    ;wait for Hi transmission
           GOTO       TESTA0H
           CLRF       TMR0       ;start timing
TESTA0L    BTFSC      PORTA,0    ;wait for Lo transmission
           GOTO       TESTA0L
           MOVF       TMR0,W     ;read value of TMR0
           SUBLW      .3         ;3-W or 3-TMR0
           BTFSC      STATUS,
                      CARRY      ;Is TMR0 > 3 i.e. a logic1
           BSF        NUMA,0     ;Yes.
```

•
•
•

This measuring of the pulse width continues until all 8 pulses are read and the relevant bits stored in the file NUMA. A TMR0 value >6 indicates the pulse was a Start pulse.

We then check to see if the number stored in the file NUMA is 27H. This is done as we have done before by subtracting 27H from it, if the answer is zero, i.e. 27−27=0, then the number transmitted was 27H and we turn on the LED. It seems such a waste to go to all this trouble to turn an LED on. I hope you can be a little more imaginative − this is only an example.

The complete codes for the transmitter and receiver are shown below as TX.ASM and RX.ASM.

The OPTION register has been set to produce timing pulses of 1ms.

Transmitter program code

TX.ASM
;tx.asm transmits code from a switch.

```
TMR0       EQU      1        ;means TMR0 is file 1.
STATUS     EQU      3        ;means STATUS is file 3.
PORTA      EQU      5        ;means PORTA is file 5.
PORTB      EQU      6        ;means PORTB is file 6.
TRISA      EQU      85H      ;TRISA (the PORTA I/O selection)
                            ;is file 85H
TRISB      EQU      86H      ;TRISB (the PORTB I/O selection)
                            ;is file 86H
OPTION_R   EQU      81H      ;the OPTION register is file 81H
ZEROBIT    EQU      2        ;means ZEROBIT is bit 2.
COUNT      EQU      0CH      ;COUNT is file 0C, a register to
                            ;count events.
NUMA       EQU      0DH
;*************************************************************

LIST       P = 16F84    ; we are using the 16F84.
ORG        0            ;the start address in memory is 0
GOTO       START        ; goto start!

;*************************************************************
;Configuration Bits

__CONFIG H'3FF0'    ;selects LP oscillator, WDT off, PUT on,
                    ; Code Protection disabled.
```

```
;********************************************************
;SUBROUTINE SECTION.


;2.5ms SECOND DELAY
DELAY1   CLRF     TMR0                    ;Start TMR0
LOOPA    MOVF     TMR0,W                  ;Read TMR0 into W
         SUBLW    .1                      ;TIME-W
         BTFSS    STATUS,ZEROBIT          ;Check TIME-W=0
         GOTO     LOOPA
         RETLW 0                          ;Return after TMR0=32


;5ms SECOND DELAY
DELAY2   CLRF     TMR0                    ;Start TMR0
LOOPB    MOVF     TMR0,W                  ;Read TMR0 into W
         SUBLW .3                         ;TIME-W
         BTFSS    STATUS,ZEROBIT          ;Check TIME-W=0
         GOTO     LOOPB
         RETLW    0                       ;Return after TMR0=2


;7.5ms SECOND DELAY
DELAY3   CLRF     TMR0                    ;Start TMR0
LOOPC    MOVF     TMR0,W                  ;Read TMR0 into W
         SUBLW .6                         ;TIME-W
         BTFSS    STATUS,ZEROBIT          ;CHECK TIME-W=0
         GOTO     LOOPC
         RETLW    0                       ;Return after TMR0=3


;********************************************************
;CONFIGURATION SECTION


START    BSF      STATUS,5       ;Turns to Bank1.
         MOVLW    B'00011111'    ;5bits of PORTA are I/P
         MOVWF    TRISA
         MOVLW    B'00000000'
         MOVWF    TRISB          ;PORTB is OUTPUT
         MOVLW    B'00000010'    ;Prescaler is /256
         MOVWF    OPTION_R       ;PRESCALER is /8,1ms


         BCF      STATUS,5       ;Return to Bank0.
         CLRF     PORTA          ;Clears PortA.
         CLRF     PORTB          ;Clears PortB.
```

```
;********************************************************
;Program starts now.


BEGIN     BTFSC      PORTA,0      ;wait for switch press
          GOTO       BEGIN
          MOVLW      27H          ;Put 27H into W
          MOVWF      NUMA         ;PUT 27H into NUMA

          BCF        PORTB,0
          CALL       DELAY1
          BSF        PORTB,0      ;Transmit START
          CALL       DELAY3       ;wait 7.5ms

TESTA0    BCF        PORTB,0      ;Transmit space
          CALL       DELAY1       ;wait 2.5ms
          BTFSC      NUMA,0       ;Test NUMA,0
          GOTO       SETA0        ;If NUMA0 = 1
          GOTO       CLRA0        ;If NUMA0 = 0

SETA0     BSF        PORTB,0      ;Transmit 1
          CALL       DELAY2       ;wait 5ms
          GOTO       TESTA1

CLRA0     BSF        PORTB,0      ;Transmit 0
          CALL       DELAY1       ;wait 2.5ms

TESTA1    BCF        PORTB,0
          CALL       DELAY1
          BTFSC      NUMA,1
          GOTO       SETA1
          GOTO       CLRA1

SETA1     BSF        PORTB,0
          CALL       DELAY2
          GOTO       TESTA2

CLRA1     BSF        PORTB,0
          CALL       DELAY1

TESTA2    BCF        PORTB,0
          CALL       DELAY1
          BTFSC      NUMA,2
          GOTO       SETA2
          GOTO       CLRA2
```

```
SETA2      BSF      PORTB,0
           CALL     DELAY2
           GOTO     TESTA3

CLRA2      BSF      PORTB,0
           CALL     DELAY1

TESTA3     BCF      PORTB,0
           CALL     DELAY1
           BTFSC    NUMA,3
           GOTO     SETA3
           GOTO     CLRA3

SETA3      BSF      PORTB,0
           CALL     DELAY2
           GOTO     TESTA4

CLRA3      BSF      PORTB,0
           CALL     DELAY1

TESTA4     BCF      PORTB,0
           CALL     DELAY1
           BTFSC    NUMA,4
           GOTO     SETA4
           GOTO     CLRA4

SETA4      BSF      PORTB,0
           CALL     DELAY2
           GOTO     TESTA5

CLRA4      BSF      PORTB,0
           CALL     DELAY1

TESTA5     BCF      PORTB,0
           CALL     DELAY1
           BTFSC    NUMA,5
           GOTO     SETA5
           GOTO     CLRA5

SETA5      BSF      PORTB,0
           CALL     DELAY2
           GOTO     TESTA6
```

```
CLRA5      BSF        PORTB,0
           CALL       DELAY1

TESTA6     BCF        PORTB,0
           CALL       DELAY1
           BTFSC      NUMA,6
           GOTO       SETA6
           GOTO       CLRA6

SETA6      BSF        PORTB,0
           CALL       DELAY2
           GOTO       TESTA7

CLRA6      BSF        PORTB,0
           CALL       DELAY1

TESTA7     BCF        PORTB,0
           CALL       DELAY1
           BTFSC      NUMA,7
           GOTO       SETA7
           GOTO       CLRA7

SETA7      BSF        PORTB,0
           CALL       DELAY2
           CLRF       PORTB
           GOTO       BEGIN

CLRA7      BSF        PORTB,0
           CALL       DELAY1
           CLRF       PORTB
           GOTO       BEGIN
END
```

Receiver program code:

```
;RX.ASM

TMR0       EQU   1      ;means TMR0 is file 1.
STATUS     EQU   3      ;means STATUS is file 3.
PORTA      EQU   5      ;means PORTA is file 5.
PORTB      EQU   6      ;means PORTB is file 6.
TRISA      EQU   85H    ;TRISA (the PORTA I/O selection) is file 85H
TRISB      EQU   86H    ;TRISB (the PORTB I/O selection) is file 86H
OPTION_R   EQU   81H    ;the OPTION register is file 81H
```

```
ZEROBIT    EQU    2          ;means ZEROBIT is bit 2.
CARRY      EQU    0
COUNT      EQU    0CH        ;COUNT is file 0C, a register to count events.
NUMA       EQU    0DH
;********************************************************

LIST       P = 16F84        ;we are using the 16F84.
ORG        0                ;the start address in memory is 0
GOTO       START            ;goto start!


;********************************************************
;Configuration Bits

__CONFIG H'3FF0'    ;selects LP oscillator, WDT off, PUT on,
                    ;Code Protection disabled.


;********************************************************
;CONFIGURATION SECTION.

START    BSF          STATUS,5       ;Turns to Bank1.

         MOVLW        B'00011111'    ;5bits of PORTA are I/P
         MOVWF        TRISA
         MOVLW        B'00000000'
         MOVWF        TRISB          ;PORTB is OUTPUT

         MOVLW        B'00000010'    ;Prescaler is /256
         MOVWF        OPTION_R       ;PRESCALER is /8,1ms

         BCF          STATUS,5       ;Return to Bank0.
         CLRF         PORTA          ;Clears PortA.
         CLRF         PORTB          ;Clears PortB.
         BCF          STATUS,5       ;Return to BANK0
         CLRF         PORTA          ;Clears PORTA
         CLRF         PORTB          ;Clears PORTB


;********************************************************
;Program starts now.

BEGIN          CLRF         NUMA

WAITHI    BTFSS     PORTA,0                  ;Wait for HI Transmission
```

```
            GOTO       WAITHI
            CLRF       TMR0
TESTST      BTFSC      PORTA,0          ;Wait for LOW Transmission
            GOTO       TESTST           ;Test for START PULSE
            MOVF       TMR0,W
            SUBLW      .5               ;5-W or 5-TMR0
            BTFSC      STATUS,CARRY     ;SKIP IF TIME>5
            GOTO       WAITHI           ;NOT START BIT
TESTA0H     BTFSS      PORTA,0          ;wait for Hi transmission
            GOTO       TESTA0H
            CLRF       TMR0             ;start timing
TESTA0L     BTFSC      PORTA,0          ;wait for Lo transmission
            GOTO       TESTA0L
            NOP
            MOVF       TMR0,W           ;read value of TMR0
            SUBLW      .3               ;3-W or 3-TMR0
            BTFSS      STATUS,CARRY     ;Is TMR0>3 i.e. a logic1
            BSF        NUMA,0           ;Yes, 1 was transmitted.

TESTA1H     BTFSS      PORTA,0          ;Wait for pulse
            GOTO       TESTA1H
            CLRF       TMR0
TESTA1L     BTFSC      PORTA,0          ;Wait for LO.
            GOTO       TESTA1L
            NOP
            MOVF       TMR0,W
            SUBLW      .3
            BTFSS      STATUS,CARRY
            BSF        NUMA,1           ;1 was transmitted

TESTA2H     BTFSS      PORTA,0          ;Wait for pulse
            GOTO       TESTA2H
            CLRF       TMR0
TESTA2L     BTFSC      PORTA,0          ;Wait for Lo.
            GOTO       TESTA2L
            NOP
            MOVF       TMR0,W
            SUBLW      .3
            BTFSS      STATUS,CARRY
            BSF        NUMA,2           ;1 was transmitted

TESTA3H     BTFSS      PORTA,0          ;Wait for pulse
            GOTO       TESTA3H
            CLRF       TMR0
```

```
TESTA3L  BTFSC   PORTA,0           ;Wait for Lo
         GOTO    TESTA3L
         NOP
         MOVF    TMR0,W
         SUBLW   .3
         BTFSS   STATUS,CARRY
         BSF     NUMA,3            ;1 was transmitted

TESTA4H  BTFSS   PORTA,0           ;Wait for pulse
         GOTO    TESTA4H
         CLRF    TMR0
TESTA4L  BTFSC   PORTA,0           ;Wait for Lo
         GOTO    TESTA4L
         NOP
         MOVF    TMR0,W
         SUBLW   .3
         BTFSS   STATUS,CARRY
         BSF     NUMA,4            ;1 was transmitted

TESTA5H  BTFSS   PORTA,0           ;Wait for pulse
         GOTO    TESTA5H
         CLRF    TMR0
TESTA5L  BTFSC   PORTA,0           ;Wait for Lo
         GOTO    TESTA5L
         NOP
         MOVF    TMR0,W
         SUBLW   .3
         BTFSS   STATUS,CARRY
         BSF     NUMA,5            ;1 was transmitted

TESTA6H  BTFSS   PORTA,0           ;Wait for pulse
         GOTO    TESTA6H
         CLRF    TMR0
TESTA6L  BTFSC   PORTA,0           ;Wait for Lo
         GOTO    TESTA6L
         NOP
         MOVF    TMR0,W
         SUBLW   .3
         BTFSS   STATUS,CARRY
         BSF     NUMA,6            ;1 was transmitted

TESTA7H  BTFSS   PORTA,0           ;Wait for pulse
         GOTO    TESTA7H
         CLRF    TMR0
```

```
TESTA7L   BTFSC    PORTA,0              ;Wait for Lo
          GOTO     TESTA7L
          NOP
          MOVF     TMR0,W
          SUBLW    .3
          BTFSS    STATUS,CARRY
          BSF      NUMA,7               ;1 was transmitted

          MOVLW    27H
          SUBWF    NUMA,W               ;NUMA-27
          BTFSS    STATUS,ZEROBIT
          GOTO     BEGIN                ;If NUMA is not 27
          BSF      PORTB,0              ;Turn on LED.
          GOTO     BEGIN

END
```

Using the transmit and receive subroutines

The transmit and receive subroutines may seem a little complex, but all you need to do in your code is call them.

- To transmit
  Put the data you wish to transmit in the file NUMA then CALL TRANSMIT. The data in the file NUMA is transmitted.
- To receive
  CALL RECEIVE, the received data will be present in the file NUMA for you to use.

These programs have illustrated how to switch an LED on (this could be a remote control for a car burglar alarm). You may of course want to add more lines of code to be able to turn the LED off. This could be done in the receiver section by waiting for say 2 seconds and on the next transmission turn the LED off, providing of course the code was again 27H. Other codes could of course be added for other switches or keypad buttons, the possibilities are endless.

The transmitter and receiver micros could be hard wired together first to test the software without the radio link. The radio transmitter and receiver can then replace the wire to give a wireless transmission.

13
EEPROM data memory

One of the special features of the 16F84, the 16F818 and some other micros is the EEPROM Data Memory. This is a section of Memory not in the usual program memory space. It is a block of data like the user files, but unlike the user files the data in the EEPROM Data Memory is saved when the microcontroller is switched off, i.e. it is non-volatile. Suppose we were counting cars in and out of a car park and we lost the power to our circuit. If we stored the count in EEPROM then we could load our count file with this data and continue without loss of data, when the power returns.

To access the data, i.e. read and write to the EEPROM memory locations, we must of course instruct the microcontroller. There are 64 bytes of EEPROM memory on the 16F84, 128 on the 16F818 and 256 on the 16F819. So we must tell the micro which address we require and if we are reading or writing to it.

When reading we identify the address from 0 to 3Fh (for the 16F84) using the address register EEADR. The data is then available in register EEDATA. When writing to the EEPROM data memory we specify the data in the register EEDATA and the location in the register EEADR.

Two other files are used to enable the process, they are EECON1 and EECON2, two EEPROM control registers.

Register EECON1 and EECON2 have addresses 8 and 9 respectively in Bank1.

The Register EECON1 is shown below in Figure 13.1.

Bit 0,   RD is set to a 1 to perform a read. It is cleared by the micro when the read is finished.
Bit 1,   WR is set to a 1 to perform a write. It is cleared by the micro when the write is finished.
Bit 2,   WREN, WRite ENable a 1 allows the write cycle, a 0 prohibits it.

bit 7	bit 6	bit 5	bit 4	bit 3	bit 2	bit 1	bit0
EEPGD	-	-	EEIF	WRERR	WREN	WR	RD

Figure 13.1 The EECON1 register

Bit 3,   WRERR reads a 1 if a write is not completed, reads a 0 if the write is completed successfully.

Bit 4,   EEIF interrupt flag for the EEDATA it is a 1 if the write operation is completed, it reads 0 if it is not completed or not started (for the 16F84). This bit has another purpose for the 16F818. We have not used this bit in this book.

Bit 7,   EEPGD, Program/Data EEPROM Select Bit. (Not used on 16F84.) This bit allows either the program memory or the data memory to be selected. 0 selects Data, 1 selects program memory.

Example using the EEPROM

As usual, I think the best way of understanding how this memory works is to look at a simple example.

Suppose we wish to count events, people going into a building, cars going into a carpark etc. So if we loose the power to the circuit the data is still retained. The circuit for this is shown in Figure 13.2.

Switch 1 is used to simulate the counting process and the 8 LEDs on PORTB display the count in binary. (This is a good chance to practice counting in binary.) The switch of course needs de-bouncing.

Remember the idea of this circuit, we are counting events and displaying the count on PORTB. But if we loose power – when the power is re-applied we want to continue the count as if nothing had happened.

So when we switch on we must move the previous EEPROM Data into the COUNT file.

The flowchart is shown below in Figure 13.3.

Just a couple of points before we look at the program:

1. It is a good idea to make sure the EEPROM DATA MEMORY is reset at the very beginning. This can be done by writing 00h to EEPROM DATA address 00h when we blow the program into the chip – this is done with the following lines of code.

```
ORG     2100H
DE      00H
```

2100H is the address of the first EEPROM data memory file i.e. 00h.

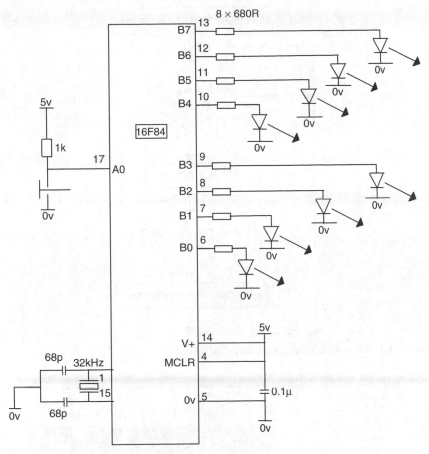

Figure 13.2 Switch press counting circuit

DE is Define EEPROM data memory, so we are initializing it with 00h, and of course 2101H is EEPROM address1 etc.

Data can also be written into the EEPROM using MPLAB, with VIEW, EEPROM and writing the data in the EEPROM box as shown in Figure 13.4.

2. Reading and Writing to EEPROM data is not as straightforward as with user files, you probably suspected that! There is a block of code you need to use – just add it to your program as required.

When reading EEPROM data at address 0 to the file COUNT then CALL READ. The subroutine written in the header.

When writing the file COUNT to EEPROM data address 0, CALL WRITE.

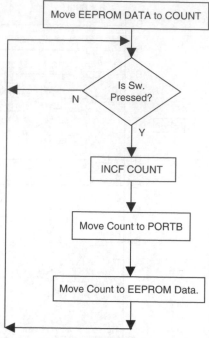

Figure 13.3 The switch press count flowchart

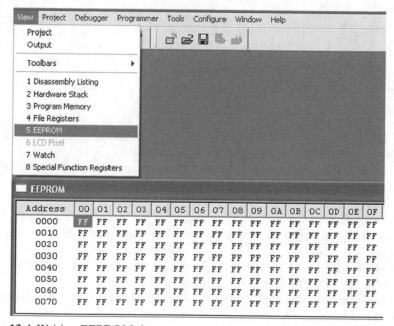

Figure 13.4 Writing EEPROM data

EEPROM program code

The complete program EEDATAWR.ASM is shown below:

```
;EEDATAWR.ASM            This program will count and display switch
;                        presses.
;                        The count is saved when the power is removed
;                        and continues when the
;                        power is re-applied.


TMR0        EQU     1           ;TMR0 is FILE 1.
PORTA       EQU     5           ;PORTA is FILE 5.
PORTB       EQU     6           ;PORTB is FILE 6.
TRISA       EQU     85H         ;TRISA (the PORTA I/O selection)
TRISB       EQU     86H         ;TRISB (the PORTB I/O selection)
OPTION_R    EQU     81H         ;the OPTION register is file 81H
STATUS      EQU     3           ;STATUS is FILE 3.
ZEROBIT     EQU     2           ;ZEROBIT is Bit 2.
COUNT       EQU     0CH         ;USER RAM LOCATION.
EEADR       EQU     9           ;EEPROM address register
EEDATA      EQU     8           ;EEPROM data register
EECON1      EQU     8           ;EEPROM control register1
EECON2      EQU     9           ;EEPROM control register2
RD          EQU     0           ;read bit in EECON1
WR          EQU     1           ;Write bit in EECON1
WREN        EQU     2           ;Write enable bit in EECON1


;********************************************************
            LIST    P=16F84     ;We are using the 16F84.
            ORG     2100H       ;ADDRESS EEADR 0
            DE      00H         ;put 00H in EEADR 0
            ORG     0           ;0 is the start address.
            GOTO    START       ;goto start!

;********************************************************
;Configuration Bits

__CONFIG H'3FF0'    ;selects LP oscillator, WDT off, PUT on,
                    ;Code Protection disabled.

;********************************************************
```

;SUBROUTINE SECTION.

;0.1 SECOND DELAY

```
DELAYP1  CLRF    TMR0           ;Start TMR0
LOOPA    MOVF    TMR0,W         ;Read TMR0 into W
         SUBLW   .3             ;TIME - W
         BTFSS   STATUS,ZEROBIT ;CHECK TIME-W = 0
         GOTO    LOOPA
         RETLW   0              ;Return after TMR0 = 3
```

;Put EEDATA 0 into COUNT

```
READ     MOVLW   0              ;read EEDATA from EEADR
                                 0 into W
         MOVWF   EEADR
         BSF     STATUS,5       ;BANK1
         BSF     EECON1,RD
         BCF     STATUS,5       ;BANK0
         MOVF    EEDATA,W
         MOVWF   COUNT
         RETLW   0
```

;WRITE COUNT INTO EEDATA 0

```
WRITE    BSF     STATUS,5       ;BANK1
         BSF     EECON1,WREN    ;set WRITE ENABLE
         BCF     STATUS,5       ;BANK0
         MOVF    COUNT,W        ;move COUNT to EEDATA
         MOVWF   EEDATA
         MOVLW   0              ;set EEADR 0 to receive
```

```
EEDATA
         MOVWF   EEADR
         BSF     STATUS,5       ;BANK1
         MOVLW   55H            ;55 and AA initiates write cycle
         MOVWF   EECON2
         MOVLW   0AAH
         MOVWF   EECON2
         BSF     EECON1,WR      ;WRITE data to EEADR 0
```

```
WRDONE   BTFSC   EECON1,WR
         GOTO    WRDONE         ;wait for write cycle to complete
```

```
            BCF        EECON1,WREN
            BCF        STATUS,5         ;BANK0
            RETLW      0
```

;**
;CONFIGURATION SECTION.

```
START       BSF        STATUS,5      ;Turn to BANK1
            MOVLW      B'00011111'   ;5 bits of PORTA are I/Ps.
            MOVWF      TRISA
            MOVLW      0
            MOVWF      TRISB         ;PORTB IS OUTPUT
            MOVLW      B'00000111'
            MOVWF      OPTION_R      ;PRESCALER is /256
            BCF        STATUS,5      ;Return to BANK0
            CLRF       PORTA         ;Clears PORTA
            CLRF       PORTB         ;Clears PORTB
            CLRF       COUNT
```

;**
;Program starts now.

```
            CALL       READ          ;read EEPROM data into COUNT
            MOVF       COUNT,W
            MOVWF      PORTB         ;Display previous COUNT (if any)
PRESS       BTFSC      PORTA,0       ;wait for switch press
            GOTO       PRESS
            CALL       DELAYP1       ;antibounce
RELEASE     BTFSS      PORTA,0       ;wait for switch release
            GOTO       RELEASE
            CALL       DELAYP1       ;antibounce

            INCF       COUNT         ;add 1 to COUNT
            MOVF       COUNT,W       ;put COUNT into W
            MOVWF      PORTB         ;move W (COUNT) to PORTB to
                                     display
            CALL       WRITE         ;write COUNT to EEPROM
                                     address 0
            GOTO       PRESS         ;return and wait for press
END
```

Microchip are continually expanding their range of microcontrollers and a new series of flash micros have been introduced, namely the 16F87X series which include 8k of program memory, 368 bytes of user RAM, 256 bytes of EEPROM data memory and an 8 channel 10 bit A/D converter. So now analogue measurements can be stored and saved in EEPROM Data!

14
Interrupts

New instructions used in this chapter:

• RETFIE

We all know what interrupts are and we don't like being interrupted. We are busy doing something and the phone rings or someone arrives at the door.

If we are expecting someone, we could look out of the window every now and again to see if they had arrived or we could carry on with what we are doing until the doorbell rings. These are two ways of receiving an interrupt. The first when we keep checking in software terms is called polling, the second when the bell rings is equivalent to the hardware interrupt.

We have looked at polling when we used the keypad to see if any keys had been pressed. We will now look at the interrupt generated by the hardware.

Before moving onto an example of an interrupt consider the action of the door in a washing machine. The washing cycle does not start until the door is closed, but after that the door does not take any part in the program. But what if a child opens the door, water could spill out or worse!! We need to switch off the outputs if the door is opened. To keep looking at the door at frequent intervals in the program (software polling) would be very tedious indeed, so we use a hardware interrupt. We carry on with the program and ignore the door. But if the door is opened the interrupt switches off the outputs – spin motor etc. If the door had been opened accidentally then closing the door would return back to the program for the cycle to continue.

This suggests that when an interrupt occurs we need to remember what the contents of the files were. i.e. the STATUS register, W register, TMR0 and PORT settings so that when we return from the interrupt the settings are restored. If we did not remember the settings, we could not continue where we left off, because the interrupt switches off all the outputs and the W register would also be altered, at the very least.

Interrupt sources

The 16F84 has 4 interrupt sources.

- Change of rising or falling edge of PORTB,0.
- TMR0 overflowing from FFh to 00h.
- PORTB bits 4–7 changing.
- DATA EEPROM write complete.

The 16F818/9 has 9 interrupt sources, and of course need extra bits in the interrupt registers to handle them. The additional interrups used in the 16F818/9 are

- A/D conversion complete
- Synchronous Serial Port Interrupt
- TMR1 overflowing
- TMR2 overflowing
- Capture Compare Pulse Width Modulator Interrupt.

These interrupts can be enabled or disabled as required by their own interrupt enable/disable bits. These bits can be found in the interrupt control register INTCON for the 16F84 and also on the Peripheral Interrupt Enable Register1, PIE1 on the 16F818/9.

In this section we will be looking at the interrupt caused by a rising or falling edge on PORTB,0.

Interrupt control register

The Interrupt Control Register INTCON, file 0Bh is shown in Figure 14.1.

bit 7	bit 6	bit 5	bit 4	bit 3	bit 2	bit 1	bit0
GIE	EEIE	T0IE	INTE	RBIE	T0IF	INTF	RBIF

Figure 14.1 The interrupt control register, INTCON of 16F84

Bit 6 in this register is designated as the Peripheral Interrupt Enable Bit, PEIE for the 16F818/9.

Before any of the individual enable bits can be switched ON, the Global Interrupt Enable (GIE) bit 7 must be set, i.e. a 1 enables all unmasked interrupts and a 0 disables all interrupts.

Bit 6   EEIE (16F84) is an EEPROM data write complete interrupt enable bit, a 1 enables this interrupt and a 0 disables it.

Bit 6   PEIE (16F818/9) is the bit that permits enabling of the extra, peripheral bits.

Bit 5   T0IE is the TMR0 overflow interrupt enable bit, a 1 enables this interrupt and a 0 disables it.

Bit 4   INTE is the RB0/INT Interrupt Enable bit, a 1 enables this interrupt and a 0 disables it.

Bit 3   RBIE is the RB Port change (B4-B7) Interrupt enable bit, a 1 enables it and a 0 disables it.

Bit 2   T0IF is the flag, which indicates TMR0 has overflowed to generate the interrupt. 1 indicates TMR0 has overflowed, 0 indicates it hasn't. This bit must be cleared in software.

Bit 1   INTF is the RB0/INT Interrupt flag bit which indicates a change on PORTB,0. A 1indicates a change has occurred, a 0 indicates it hasn't.

Bit 0   RBIF is the RB PORT Change Interrupt flag bit. A 1 indicates that one of the inputs PORTB,4–7 has changed state. This bit must be cleared in software. A 0 indicates that none of the PORTB,4-7 bits have changed.

Program using an interrupt

As an example of how an interrupt works consider the following example:

Suppose we have 4 lights flashing consecutively for 5 seconds each. A switch connected to B0 acts as an interrupt so that when B0 is at a logic 0 an interrupt routine is called. This interrupt routine flashes all 4 lights ON and OFF twice at 1 second intervals and then returns back to the program providing the switch on B0 is at a logic1.

I have used the 16F818 for this application.

The circuit diagram for this application is shown in Figure 14.2.

One thing to note from the circuit the 16F818 chip has internal pull-up resistors on PORTB so B0 does not need a pull up resistor on the switch.

The interrupt we are using is a change on B0, we are therefore concerned with the following bits in the INTCON register, i.e. INTE bit4 the enable bit and INTF bit1 the flag showing B0 has changed, and of course GIE bit7 the Global Interrupt Enable Bit.

Program operation

When B0 generates an interrupt the program branches to the interrupt service routine. Where? Program memory location 4 tells the Microcontroller where to go to find the interrupt service routine.

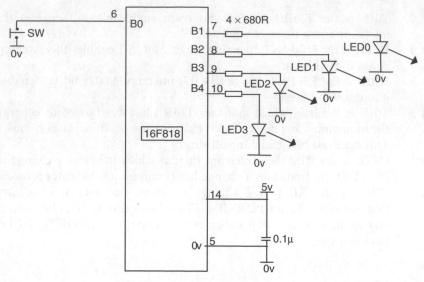

Figure 14.2 Interrupt demonstration circuit

Program memory location 4 is then programmed using the org statement as:

```
ORG     4           ;write next instruction in program memory location 4
GOTO    ISR         ;jump to the Interrupt Service Routine.
```

The interrupt service routine

The Interrupt Service Routine, ISR, is written like a subroutine and is shown below:

```
;Interrupt Service Routine
        MOVWF       W_TEMP          ;Save W
        SWAPF       STATUS,W
        MOVWF       STATUS_T        ;Save STATUS
        MOVF        TMR0,W
        MOVWF       TMR0_T          ;Save TMR0
        MOVF        PORTB,W
        MOVWF       PORTB_T         ;Save PORTB

        MOVLW       0FFH
        MOVWF       PORTB           ;turn on all outputs.
        CALL        DELAPY1         ;1 second delay
```

```
            MOVLW      0
            MOVWF      PORTB          ;turn off all outputs
            CALL       DELAPY1        ;1 second delay
            MOVLW      0FFH
            MOVWF      PORTB          ;turn on all outputs.
            CALL       DELAPY1        ;1 second delay
            MOVLW      0
            MOVWF      PORTB          ;turn off all outputs
            CALL       DELAPY1        ;1 second delay
SW_HI       BTFSS      PORTB,0
            GOTO       SW_HI          ;wait for switch to be HI.


            SWAPF      STATUS_T,W
            MOVWF      STATUS         ;Restore STATUS
            MOVF       TMR0_T,W
            MOVWF      TMR0           ;Restore TMR0
            MOVF       PORTB_T,W
            MOVWF      PORTB          ;Restore PORTB
            MOVF       W_TEMP,W       ;Restore W


            BCF        INTCON, INTF   ;Reset Interrupt Flag
            RETFIE                    ;Return from the interrupt
```

Operation of the interrupt service routine

The interrupt service routine operates in the following way.

- When an interrupt is made the Global Interrupt Enable is cleared automatically (disabled) to switch off all further interrupts. We would not wish to be interrupted while we are being interrupted.
- The registers W, STATUS, TMR0 and PORTB are saved in temporary locations W_TEMP, STATUS_T, TMR0_T and PORTB_T.
- The interrupt routine is executed, the lights flash on and off twice. This is a separate sequence than before to show the interrupt has interrupted the normal flow of the program. NB. The program has not been looking at the switch that generated the interrupt.
- We then wait until the switch returns HI.
- The temporary files W_TEMP, STATUS_T, TMR0_T and PORTB_T are restored back into W, STATUS, TMR0 and PORTB.
- The PORTB,0 interrupt flag INTCON,INTF is cleared ready to indicate further interrupts.
- We return from the interrupt, and the Global Interrupt Enable bit is automatically set to enable further interrupts.

Program of the interrupt demonstration

The complete code for this program is shown below as INTFLASH.ASM.

```
;INTFLASH.ASM Flashing lights being interrupted by a switch on B0.
;Using 16F818
;EQUATES SECTION


TMR0        EQU    1        ;means TMR0 is file 1.
STATUS      EQU    3        ;means STATUS is file 3.
PORTA       EQU    5        ;means PORTA is file 5.
PORTB       EQU    6        ;means PORTB is file 6.
TRISA       EQU    85H      ;TRISA (the PORTA I/O selection) is file 85H
TRISB       EQU    86H      ;TRISB (the PORTB I/O selection) is file 86H
INTCON      EQU    0BH      ;Interrupt Control Register
ZEROBIT     EQU    2        ;means ZEROBIT is bit 2.
CARRY       EQU    0        ;CARRY IS BIT 0.
GIE         EQU    7        ;Global Interrupt bit
INTE        EQU    4        ;B0 interrupt enable bit.
INTF        EQU    1        ;B0 interrupt flag
OPTION_R    EQU    81H
ADCON0      EQU    1FH      ;A/D Configuration reg.0
ADCON1      EQU    9FH      ;A/D Configuration reg.1
ADRES       EQU    1EH      ;A/D Result register.
OSCCON      EQU    8FH      ;Oscillator control register.
COUNT       EQU    20H      ;COUNT a register to count events.
                           ;a register to count events
TMR0_T      EQU    21H      ;TMR0 temporary file
W_TEMP      EQU    22H      ;W temporary file
STATUS_T    EQU    23H      ;STATUS temporary file
PORTB_T     EQU    24H      ;PORTB temporary file
COUNTA      EQU    25H
;****************************************************************


LIST        P=16F818               ;we are using the 16F818.
            ORG      0             ;the start address in memory is 0
            GOTO     START         ;goto start!
            ORG      4             ;write to memory location 4
            GOTO     ISR           ;location4 jumps to ISR
;****************************************************************


;Configuration Bits
__CONFIG H'3F10'           ;sets INTRC-A6 is port I/O, WDT off, PUT
```

```
                              ;on, MCLR tied to VDD A5 is I/O
                              ;BOD off, LVP disabled, EE protect disabled,
                              ;Flash Program Write disabled,
                              ;Background Debugger Mode disabled, CCP
                              ;function on B2,
                              ;Code Protection disabled.
;***********************************************************

;SUBROUTINE SECTION

;0.1 second delay, actually 0.099968s
DELAYP1  CLRF     TMR0              ;START TMR0.
LOOPB    MOVF     TMR0,W            ;READ TMR0 INTO W.
         SUBLW    .3                ;TIME-3
         BTFSS    STATUS,ZEROBIT    ;Check TIME-W = 0
         GOTO     LOOPB             ;Time is not = 3.
         NOP                        ;add extra delay
         NOP
         RETLW    0                 ;Time is 3, return.


;5 second delay.
DELAY5   MOVLW    .50
         MOVWF    COUNTA
LOOPC    CALL     DELAYP1
         DECFSZ   COUNTA
         GOTO     LOOPC
         RETLW    0


;1 second delay.
DELAY1   MOVLW    .10
         MOVWF    COUNT
LOOPA    CALL     DELAYP1
         DECFSZ   COUNT
         GOTO     LOOPA
         RETLW    0

;Interrupt Service Routine.

ISR      MOVWF    W_TEMP          ;Save W
         SWAPF    STATUS,W
         MOVWF    STATUS_T        ;Save STATUS
         MOVF     TMR0,W
         MOVWF    TMR0_T          ;Save TMR0
         MOVF     PORTB,W
         MOVWF    PORTB_T         ;Save PORTB
```

```
            MOVLW        0FFH
            MOVWF        PORTB           ;turn on all outputs.
            CALL         DELAY1          ;1 second delay
            MOVLW        0
            MOVWF        PORTB           ;turn off all outputs
            CALL         DELAY1          ;1 second delay
            MOVLW        0FFH
            MOVWF        PORTB           ;turn on all outputs.
            CALL         DELAY1          ;1 second delay
            MOVLW        0
            MOVWF        PORTB           ;turn off all outputs
            CALL         DELAY1          ;1 second delay
SW_HI       BTFSS        PORTB,0
            GOTO         SW_HI           ;wait for switch to be HI.

            SWAPF        STATUS_T,W
            MOVWF        STATUS          ;Restore STATUS
            MOVF         TMR0_T,W
            MOVWF        TMR0            ;Restore TMR0
            MOVF         PORTB_T,W
            MOVWF        PORTB           ;Restore PORTB
            MOVF         W_TEMP,W        ;Restore W

            BCF          INTCON,INTF     ;Reset Interrupt Flag
            RETFIE                       ;Return from the interrupt
```

```
;********************************************************
;CONFIGURATION SECTION
```

```
START       BSF          STATUS,5        ;Turns to Bank1.
            MOVLW        B'11111111'     ;8 bits of PORTA are I/P
            MOVWF        TRISA

            MOVLW        B'00000110'     ;PORTA IS DIGITAL
            MOVWF        ADCON1

            MOVLW        B'00000001'
            MOVWF        TRISB           ;PORTB,0 is I/P

            MOVLW        B'00000000'
            MOVWF        OSCCON          ;oscillator 31.25kHz

            MOVLW        B'00000111'     ;Prescaler is /256
            MOVWF        OPTION_R        ;TIMER is 1/32 secs.
```

```
              BCF          STATUS,5       ;Return to Bank0.
              CLRF         PORTA          ;Clears PortA.
              CLRF         PORTB          ;Clears PortB.
              BSF          INTCON,GIE     ;Enable Global Interrupt
              BSF          INTCON,INTE    ;Enable B0 interrupt
```

;**
;
;Program starts now.

```
BEGIN         MOVLW        B'00000010'    ;Turn on B1
              MOVWF        PORTB
              CALL         DELAY5         ;wait 5 seconds
              MOVLW        B'00000100'    ;Turn on B2
              MOVWF        PORTB
              CALL         DELAY5         ;wait 5 seconds
              MOVLW        B'00001000'    ;Turn on B3
              MOVWF        PORTB
              CALL         DELAY5         ;wait 5 seconds
              MOVLW        B'00010000'    ;Turn on B4
              MOVWF        PORTB
              CALL         DELAY5         ;wait 5 seconds
              GOTO         BEGIN
END
```

The 4 lights are flashing on and off slowly enough (5 second intervals) so that you can interrupt part way through taking B0 low via the switch, (make sure B0 is hi when starting). The interrupt service routine then flashes all the lights on and off twice at 1 second intervals.

When returning from the interrupt with B0 hi again, the program resumes from where it left off, i.e. if the 2nd LED had been on for 3 seconds it would come back on for the remaining 2 seconds and the sequence would continue.

15
The 12 series 8 pin microcontroller

Arizona Microchip have a range of microcontrollers with 8 pins. They include types with Data EEPROM and A/D converters. In this section we will cover the 12C508 and 12C509, which are one time programmable devices and the flash 12F629 and 12F675 (electronically) reprogrammable devices.

The device memory specifications are shown in Table 15.1.

Table 15.1 12C508/509, 12F629 and 12F675 memory specifications

Device	EEPROM	User Files	Registers
12C508	512×12	25	7
12C509	1024×12	41	7
12F629	1024×14	64	29
12F675	1024×14	64	33

Pin diagram of the 12C508/509

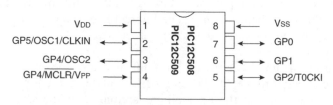

Figure 15.1 Pin diagram of the 12C508/9

Pin diagram of the 12F629 and 12F675

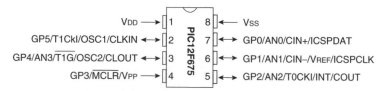

Figure 15.2 Pin diagram of the 12F629 and 12F675

Features of these 12 series

One of the special features of this Micro is that it has 8 pins, but 6 of them can be used as I/O pins, the remaining 2 pins being used for the power supply. There is no need to add a crystal and capacitors, because a 4MHz oscillator is built on board! If you wish to use a clock other than the 4MHz provided, then you can connect an oscillator circuit to pins 2 and 3 (as in the 16F84). That leaves you with of course only 4 I/O.

Being an 8 pin device means of course it is smaller than an 18 pin device and cheaper. The on board oscillator means that the crystal and timing capacitors are not required, reducing the component count, size and cost even further. So if your application requires no more than 6 I/O these are devices to use. They have useful applications in burglar alarm circuits and the radio transmitter circuits we have looked at previously.

The memory maps of the 12C508 and 12F629/675

The memory map of the 12C508 is shown in Figure 15.3, showing the 7 registers and 25 user files. Figure 15.4 shows the 12F629/675 map.

The 12C509 has 16 extra user files mapped in Bank1.

There is no longer a PORTA or PORTB because we only have 6 I/O, they are in a port called GPIO (General Purpose Input Output), File 6.

Address	File
01h	TMR0
02h	PCL
03h	STATUS
04h	FSR
05h	OSCCAL
06h	GPIO
07h	
	General Purpose Registers (User files)
1Fh	

Figure 15.3 12C508 Memory map

Address	Register
00H	INDADRESS
01H	TMR0
02H	PCL
03H	STATUS
04H	FSR
05H	GPIO
06H	
07H	
08H	
09H	
0AH	PCLATH
0BH	INTCON
0CH	PIR1
0DH	
0EH	TMR1L
0FH	TMR1H
10H	T1CON
11H	
12H	
13H	
14H	
15H	
16H	
17H	
18H	
19H	CMCON
1AH	
1BH	
1CH	
1DH	
1EH	ADRESH
1FH	ADRESL
20H	General Purpose Register
5FH	64 bytes

BANK 0

Address	Register
80H	INDADRR
81H	OPTION REG
82H	PCL
83H	STATUS
84H	FSR
85H	TRISIO
86H	
87H	
88H	
89H	
8AH	PCLATH
8BH	INTCON
8CH	PIE1
8DH	
8EH	PCON
8FH	
90H	OSCCAL
91H	
92H	
93H	
94H	
95H	WPU
96H	IOCB
97H	
98H	
99H	VRCON
9AH	EEDATA
9BH	EEADR
9CH	EECON1
9DH	EECON2
9EH	ADRESL
9FH	ANSEL

BANK 1

Figure 15.4 12F629/675 Memory map

Oscillator calibration

Apart from the small size of this device an appealing feature is that the oscillator is on board. The file OSCCAL is an oscillator calibration file used to trim the 4MHz oscillator.

The 4MHz oscillator takes its timing from an on board R-C network, which is not very precise. So these chips have a value that can be put into OSCCAL to trim it. This value is stored in the last memory address i.e. 01FFh for the 12C508 and 03FFh for the 12C509 and 12F629/675.

- Trimming the 12C508/9

The code, which is loaded by the manufacturer in the last memory location for the 12C508/9, is MOVLW XX where XX is the trimming value. The last memory location is the reset vector i.e. when switched on the micro goes to this location first, it loads the calibration value into W and the program counter overflows to 000h and continues executing the code. To use the calibration value, in the Configuration Section write the instruction MOVWF OSCCAL, which then moves the manufacturers calibration value into the timing circuit.

There is one point to remember – if you are using a windowed device then the calibration value will be erased when the memory is erased. So make a note of the MOVLW XX code by looking in MPLAB with: VIEW-PROGRAM MEMORY and program it back in by ORG 01FFH MOVLW XX.

- Trimming the 12F629/675

A calibration instruction is programmed into the last location of program memory, i.e. 3FFH. The instruction is RETLW XX, where XX is the calibration value. This value is placed in the OSCCAL register to set the calibration value of the internal oscillator. This is done in the 12F629 header as

```
CALL    3FFH        ;call instruction at location 3FFH
MOVWF  OSCCAL      ;move calibration value to OSCCAL
```

The trimming can be ignored if required – but it only requires 1 or 2 lines of code, so why not use it.

I/O PORT, GPIO

The GPIO, General Purpose Input/Output, is an 8 bit I/O register, it has 6 I/O lines available so bits GPIO 0 to 5 are used, bits 6 and 7 are not.
N.B. GPIO bit3 is an input only pin so there is a maximum of 5 outputs.

- For the 12C508 GPIO pins 0,1 and 3 can be configured with weak pull ups by writing 0 to OPTION,6 (bit 6 in the OPTION register).
- For the 12F629/675 all GPIO pins except GPIO3 can be configured with weak pull ups. This is done by setting the relevant bits in the Weak Pull Up Register, WPU. When in

```
Bank1      MOVLW    B'00110111'
           MOVWF    WPU
```

Will turn on all the weak pull ups.

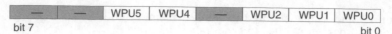

bit 7   bit 0

Figure 15.5 Weak pull up register

Delays with the 12 series

We have previously used a 32kHz. Crystal with the 16F84 device, but now we are going to use the internal 4MHz clock.

A 4MHz clock means that the basic timing is ¼ of this i.e. 1MHz. If we set the OPTION register to divide by 256 this gives a timing frequency of 3906Hz. In the headers for the 12C508/9, 12F629 and 12F675 I have (as with the 16F84) included a one second and a 0.5 second delay. In order to achieve a one second delay from a frequency of 3906Hz I first of all produced a delay of 1/100 second by counting 39 timing pulses i.e. 3906Hz/39 = 100.15 = 100Hz approx., called DELAY. A one second delay, subroutine DELAY1 then counts 100 of these DELAY times (i.e. 100 × 1/100 second), and of course a delay of 0.5 seconds would count 50.

Just before we look at the headers – we do not have an instruction SUBLW on the 12C508. I have therefore set up a file called TIME that I have written 39 into. I then move TMR0 into W and subtract the file TIME (39d) from it to see if TMR0 = 39 i.e. 1/100 of a second has elapsed.

WARNING: The 12C508 and 509 micros only have a two level deep stack. Which means when you do e.g. a one second delay, CALL DELAY1 this then calls another subroutine, i.e. CALL DELAY. You have used your two levels and cannot do any further calls without returning from one at least one of those subroutines. If you did make a third CALL the program would not be able to find its way back!

Header for 12C508/9

```
;HEAD12C508.ASM FOR 12C508/9.

TMR0      EQU      1          ;TMR0 is FILE 1.
OSCCAL    EQU      5          ;Oscillator calibration
GPIO      EQU      6          ;GPIO is FILE 6.
STATUS    EQU      3          ;STATUS is FILE 3.
ZEROBIT   EQU      2          ;ZEROBIT is Bit 2.
COUNT     EQU      07H        ;USER RAM LOCATION.
TIME      EQU      08H        ;TIME IS 39
;****************************************************************
```

```
LIST      P=12C508    ;We are using the 12C508.
ORG       0           ;0 is the start address.
GOTO      START       ;goto start!


;*****************************************************
Configuration Bits

__CONFIG H'0FEA'     ;selects internal RC oscillator, WDT off,
                     ;code protection disabled
;*****************************************************

;SUBROUTINE SECTION.

;1/100 SECOND DELAY
DELAY     CLRF    TMR0                  ;Start TMR0
LOOPA     MOVF    TMR0,W                ;Read TMR0 into W
          SUBWF   TIME,W                ;TIME-W
          BTFSS   STATUS,ZEROBIT  ;Check TIME-W=0
          GOTO    LOOPA
          RETLW   0                     ;Return after TMR0 = 39

;1 SECOND DELAY
DELAY1 MOVLW    .100
          MOVWF   COUNT
TIMEA     CALL    DELAY
          DECFSZ  COUNT
          GOTO    TIMEA
          RETLW   0

;1/2 SECOND DELAY
DELAYP5   MOVLW   .50
          MOVWF   COUNT
TIMEB     CALL    DELAY
          DECFSZ  COUNT
          GOTO    TIMEB
          RETLW   0


;*****************************************************
; CONFIGURATION SECTION.

START     MOVWF   OSCCAL        ;Calibrate oscillator.
          MOVLW   B'00001000'   ;5 bits of GPIO are O/Ps.
          TRIS    GPIO          ;Bit3 is Input
          MOVLW   B'00000111'
```

```
OPTION                    ;PRESCALER is /256
CLRF       GPIO           ;Clear GPIO
MOVLW      .39
MOVWF      TIME           ;TIME = 39
```

```
;****************************************************************
;Program starts now.
```

Program application for 12C508

There are 5 I/O on the 12C508 i.e. GPIO bits 0,1,2,4 and 5. Bit3 is an input only. For our application we will chase 5 LEDs on our outputs backwards and forwards at 0.5 second intervals.

The Circuit diagram is shown in Figure 15.6.

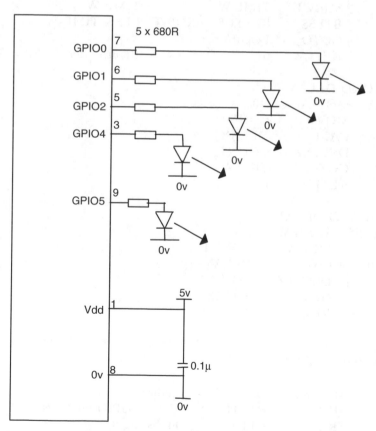

Figure 15.6 LED chasing circuit for the 12C508

Notice that the only other component required is the power supply decoupling capacitor, 0.1µF, no oscillator circuit is required.

The program for the LED Chasing Project, LED_CH12.ASM is shown below.

;LED_CH12.ASM Program to chase 5 LEDs with the 12C508

```
TMR0       EQU      1            ;TMR0 is FILE 1.
OSCCAL     EQU      5
GPIO       EQU      6            ;GPIO is FILE 6.
STATUS     EQU      3            ;STATUS is FILE 3.
ZEROBIT    EQU      2            ;ZEROBIT is Bit 2.
COUNT      EQU      07H          ;USER RAM LOCATION.
TIME       EQU      08H          ;TIME IS 39
;************************************************************
LIST       P=12C508              ;We are using the 12C508.
ORG        0                     ;0 is the start address.
GOTO       START                 ;goto start!


;************************************************************
;Configuration Bits

_CONFIG H'0FEA'         ;selects Internal RC oscillator, WDT off,
                        ;Code Protection disabled.


;************************************************************
;SUBROUTINE SECTION.

DELAY      CLRF     TMR0                    ;Start TMR0
LOOPA      MOVF     TMR0,W                  ;Read TMR0 into W
           SUBWF    TIME,W                  ;TIME - W
           BTFSS    STATUS,ZEROBIT          ;Check TIME-W=0
           GOTO     LOOPA
           RETLW    0                       ;Return after TMR0 = 39

;1 SECOND DELAY
DELAY1     MOVLW    .100
           MOVWF    COUNT
TIMEA      CALL     DELAY
           DECFSZ   COUNT
           GOTO     TIMEA
           RETLW    0

;1/2 SECOND DELAY
DELAYP5    MOVLW    .50
           MOVWF    COUNT
```

```
TIMEB    CALL      DELAY
         DECFSZ    COUNT
         GOTO      TIMEB
         RETLW     0
```

```
;**********************************************************
;CONFIGURATION SECTION.
```

```
START    MOVWF     OSCCAL     ;Calibrate oscillator.

         MOVLW     B'00001000'  ;5 bits of GPIO are O/Ps.
         TRIS      GPIO         ;Bit3 is Input
         MOVLW     B'00000111'
         OPTION                 ;PRESCALER is /256
         CLRF      GPIO         ;Clear GPIO
         MOVLW     .39
         MOVWF     TIME         ;TIME = 39
```

```
;**********************************************************
;Program starts now.
```

```
BEGIN    MOVLW     B'00000001'  ;turn on LED0
         MOVWF     GPIO
         CALL      DELAYP5
         MOVLW     B'00000010'  ;turn on LED1
         MOVWF     GPIO
         CALL      DELAYP5
         MOVLW     B'00000100'  ;turn on LED2
         MOVWF     GPIO
         CALL      DELAYP5
         MOVLW     B'00010000'  ;turn on LED3
         MOVWF     GPIO
         CALL      DELAYP5
         MOVLW     B'00100000'  ;turn on LED4
         MOVWF     GPIO
         CALL      DELAYP5
         MOVLW     B'00010000'  ;turn on LED3
         MOVWF     GPIO
         CALL      DELAYP5
         MOVLW     B'00000100'  ;turn on LED2
         MOVWF     GPIO
         CALL      DELAYP5
         MOVLW     B'00000010'  ;turn on LED1
```

```
            MOVWF    GPIO
            CALL     DELAYP5
            GOTO     BEGIN
END
```

The program is similar in content to the 16F84 programs used previously, but with the following exceptions:

- A file TIME, file 8, has been set up which has had 39 loaded into it, in the Configuration Section. This is used to determine when TMR0 has reached a count of 39, time of 0.01 seconds, which is then used in the timing subroutines.
- In the Configuration Section the first instruction the program encounters is MOVWF OSCCAL. This moves the calibration value which has just been read by MOVLW XX, from location 1FFH, the first instruction, into the calibration file OSCCAL.
- GPIO is used in the program instead of the usual PORTA and PORTB.

Program application using the 12F629/675

To perform the LED chasing action of the previous example in Figure 15.6 using the 12F675 the following code would be required.

;LED_CH675.ASM FOR 12F675 using 4MHz internal RC.

```
TMR0       EQU    1         ;TMR0 is FILE 1.
TRISIO     EQU    85H
GPIO       EQU    5         ;GPIO is FILE 6.
STATUS     EQU    3         ;STATUS is FILE 3.
ZEROBIT    EQU    2         ;ZEROBIT is Bit 2.
GO         EQU    1
ADSEL      EQU    9EH
ADCON0     EQU    1FH
ADRESH     EQU    1EH
OPTION_R   EQU    81H
CMCON      EQU    19H
OSCCAL     EQU    90H
COUNT      EQU    20H       ;USER RAM LOCATION.

;***************************************************
;
LIST       P=12F675         ;We are using the 12F675.
ORG        0                ;0 is the start address.
GOTO       START            ;goto start!
```

```
;**************************************************
;Configuration Bits

__CONFIG H'3F84'          ;selects Internal RC oscillator, WDT off,
                         ;Code Protection disabled.

;******************************************************
;SUBROUTINE SECTION.

;1/100 SECOND DELAY
DELAY    CLRF    TMR0                 ;START TMR0
LOOPA    MOVF    TMR0,W               ;READ TMR0 IN W
         SUBLW   .39                  ;TIME-W
         BTFSS   STATUS,ZEROBIT       ;CHECK TIME-W=0
         GOTO    LOOPA
         RETLW   0                    ;RETURN AFTER TMR0 = 39

;P1 SECOND DELAY
DELAYP1  MOVLW      .10
         MOVWF      COUNT
TIMEC    CALL       DELAY
         DECFSZ     COUNT
         GOTO       TIMEC
         RETLW      0

;P5 SECOND DELAY
DELAYP5  MOVLW      .50
         MOVWF      COUNT
TIMED    CALL       DELAY
         DECFSZ     COUNT
         GOTO       TIMED
         RETLW      0

;**************************************************
;CONFIGURATION SECTION.

START    BSF        STATUS,5      ;BANK1
         MOVLW      B'00010000'   ;All I/O are digital (12F675 only)
         MOVWF      ADSEL

         MOVLW      B'00001000'   ;Bit3 is IP
         MOVWF      TRISIO

         MOVLW      B'00000111'
         MOVWF      OPTION_R      ;PRESCALER is /256
```

```
                CALL        3FFH
                MOVWF       OSCCAL      ;Calibrates 4MHz oscillator

                BCF         STATUS,5    ;BANK0

                MOVLW       7H
                MOVWF       CMCON       ;Turns off comparator
                CLRF        GPIO        ;Clears GPIO
                BSF         ADCON0,0    ;Turns on A/D converter.


;***********************************************************
;
;Program starts now.
BEGIN           MOVLW       B'00000001' ;turn on LED0
                MOVWF       GPIO
                CALL        DELAYP5
                MOVLW       B'00000010' ;turn on LED1
                MOVWF       GPIO
                CALL        DELAYP5
                MOVLW       B'00000100' ;turn on LED2
                MOVWF       GPIO
                CALL        DELAYP5
                MOVLW       B'00010000' ;turn on LED3
                MOVWF       GPIO
                CALL        DELAYP5
                MOVLW       B'00100000' ;turn on LED4
                MOVWF       GPIO
                CALL        DELAYP5
                MOVLW       B'00010000' ;turn on LED3
                MOVWF       GPIO
                CALL        DELAYP5
                MOVLW       B'00000100' ;turn on LED2
                MOVWF       GPIO
                CALL        DELAYP5
                MOVLW       B'00000010' ;turn on LED1
                MOVWF       GPIO
                CALL        DELAYP5
                GOTO        BEGIN
END
```

The differences in the code between the 12C508 and 12F675 are:

- MOVLW   B'00010000' ;All I/O are digital (12F675 only)
  MOVWF   ADSEL

These two lines are used to inform the 12F675 that the inputs are all digital. Change the data to make the inputs analogue – refer to manufacturers data. These two lines are not required for the 12F629 which does not have any A/D.

- CALL   3FFH
  MOVWF   OSCCAL   ;Calibrates 4MHz oscillator

These lines are used to calibrate the internal 4MHz oscillator.

- MOVLW   7H
  MOVWF   CMCON   ;Turns off comparator

The 12F629/675 have analogue comparators, which we have not looked at. They need to be turned off to use the I/O pins. The default is that the comparators are on!

There are numerous other 12 series microcontrollers but once you have understood how to move from the 12C508/9 to the 12F629/675 you will be able to migrate to the rest.

16
The 16F87X microcontroller

The 16F87X range includes the devices, 16F870, 16F871, 16F872, 16F873, 16F874, 16F876 and 16F877. They are basically the same device but differ in the amounts of I/O, analogue inputs, program memory, data memory (RAM) and EEPROM data memory that they have.

The 16F87X have more I/O, program memory, data memory, EEPROM data memory and analogue inputs than the 16F818.

16F87X family specification

Device	Program Memory	EEPROM Data Memory (bytes)	RAMBytes	Pins	I/O	10 bit A/D Channels
16F870	2k	64	128	28	22	5
16F871	2k	64	128	40	33	8
16F872	2k	64	128	28	22	5
16F873	4k	128	192	28	22	5
16F874	4k	128	192	40	33	8
16F876	8k	256	368	28	22	5
16F877	8k	256	368	40	33	8

16F87X memory map

The 16F87X devices have more functions than we have seen previously. These functions of course need registers in order to make the various selections.

The memory map of the 16F87X showing these registers is shown in Figure 16.3.

The 16F87X devices have a number of extra registers that are not required in the applications we have looked at. For an explanation of these registers please see Microchip's website @ www.microchip.com, where you can download the data sheet as a pdf (portable document file), which can be read using Adobe Acrobat Reader.

MCLR/Vpp/THV	1	28	B7/PGD	
A0/AN0	2	27	B6/PGC	
A1/AN1	3	26	B5	
A2/AN2/Vref-	4	25	B4	
A3/AN3/Vref+	5	24	B3/PGM	
A4/T0CKI	6	23	B2	
A5/AN4/SS	7	22	B1	
Vss	8	21	B0/INT	
OSC1/CLKIN	9	20	Vdd	
OSC2/CLKOUT	10	19	Vss	
C0/T1OSO/T1CLKI	11	18	C7/RX/DT	
C1/T1OSI/CCP2	12	17	C6/TX/CK	
C2/CCP1	13	16	C5/SDO	
C3/SCK/SCL	14	15	C4/SD1/SDA	

Figure 16.1 The 16F870/2/3/6 pinout

The 16F872 microcontroller

In order to demonstrate the operation of the 16F87X series we will consider the 16F872 device. This is a 28pin device with 22 I/O available on 3 ports. PortA has 6 I/O, PortB has 8 I/O and PORTC has 8 I/O. Of the 6 I/O available on PortA 5 of them can be analogue inputs. The header for the 16F872, HEAD872.ASM, configures the device with 5 analogue inputs on PortA, 8 digital inputs on PortC and 8 outputs on PortB. The port configuration for the device is shown in Figure 16.4.

The 16F872 has been configured in HEAD872.ASM, using a 32 kHz crystal, to allow all the programs used previously to be copied over with as little alteration as possible.

Devices included in this Data Sheet:

- PIC16F873
- PIC16F876
- PIC16F874
- PIC16F877

Microcontroller Core Features:

- High performance RISC CPU
- Only 35 single word instructions to learn
- All single cycle instructions except for program branches which are two cycle
- Operating speed: DC - 20 MHz clock input
  DC - 200 ns instruction cycle
- Up to 8K × 14 words of FLASH Program Memory,
  Up to 368 × 8 bytes of Data Memory (RAM)
  Up to 256 × 8 bytes of EEPROM Data Memory
- Pinout compatible to the PIC16C73B/74B/76/77
- Interrupt capability (up to 14 sources)
- Eight level deep hardware stack
- Direct, indirect and relative addressing modes
- Power-on Reset (POR)
- Power-up Timer (PWRT) and
  Oscillatior Start-up Times (OST)
- Watchdog Timer (WDT) with its own on-chip RC oscillator for reliable operation
- Programmable code-protection
- Power saving SLEEP mode
- Selectable oscillator options
- Low power, high speed CMOS FLASH/EEPROM technology
- Fully static design
- In-Circuit Serial Programming™ (ICSP) via two pins
- Single 5V In-Circuit Serial Programming capability
- In-Circuit Debugging via two pins
- Processor read/write access to program memory
- Wide operating voltage range: 2.0V to 5.5V
- High Sink/Source Current: 25 mA
- Commercial and Industrial and Extended temperature ranges
- Low-power consumption:
  - < 2 mA typical @ 3V, 4 MHz
  - 20 μA typical @3V, 32 kHz
  - < 1 μA typical standby current

Pin Diagram

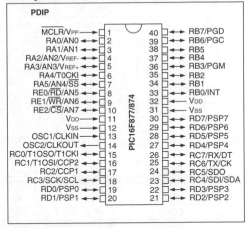

Peripheral Features:

- Timer0: 8-bit timer/counter with 8-bit prescaler
- Timer1: 16-bit timer/counter with prescaler, can be incremented during SLEEP via external crystal/clock
- Timer2: 8-bit timer/counter with 8-bit period register, rescaler and postscaler
- Two Capture, Compare, PWM modules
  - Capture is 16-bit, max. resolution is 12.5 ns
  - Compare is 16-bit, max. resolution is 200 ns
  - PWM max. resolution is 10-bit
- 10-bit multi-channel Analog-to-Digital converter
- Synchronous Serial Port (SSP) with SPI™ (Master mode) and I²C™ (Master/Slave)
- Universal Synchronous Asynchronous Receiver Transmitter (USART/SCI) with 9-bit address detection
- Parallel Slave Port (PSP) 8-bits wide, with external $\overline{RD}$, $\overline{WR}$ and $\overline{CS}$ controls (40/44-pin only)
- Brown-out detection circuitry for Brown-out Reset (BOR)

Figure 16.2 The 16F87X data sheet

Address	File Name Bank0	File Name Bank1	File Name Bank2	File Name Bank3
00h	Ind.Add	Ind.Add	Ind.Add	Ind.Add
01h	TMR0	Option	TMR0	Option
02h	PCL	PCL	PCL	PCL
03h	Status	Status	Status	Status
04h	FSR	FSR	FSR	FSR
05h	PORTA	TRISA		
06h	PORTB	TRISB	PORTB	TRISB
07h	PORTC	TRISC		
08h	PORTD	TRISD		
09h	PORTE	TRISE		
0Ah	PCLATH	PCLATH	PCLATH	PCLATH
0Bh	INTCON	INTCON	INTCON	INTCON
0Ch	PIR1	PIE1	EEDATA	EECON1
0Dh	PIR2	PIE2	EEADR	EECON2
0Eh	TMR1L	PCON	EEDATH	
0Fh	TMR1H		EEADRH	
10h	T1CON			
11h	TMR2	SSPCON2		
12h	T2CON	PR2		
13h	SSPBUF	SSPADD		
14h	SSPCON	SSPSTAT		
15h	CCPR1L			
16h	CCPR1H			
17h	CCP1CON		General Purpose Register 96 bytes	General Purpose Register 96 bytes
18h	RCSTA	TXSTA		
19h	TXREG	SPBRG		
1Ah	RCREG			
1Bh	CCPR2L			
1Ch	CCPR2H			
1Dh	CCP2CON			
1Eh	ADRESH	ADRESL		
1Fh	ADCON0	ADCON1		
.	General Purpose Register	General Purpose Register		
6Fh	96 bytes	80 bytes		
7FH				

Figure 16.3 The 16F87X memory map

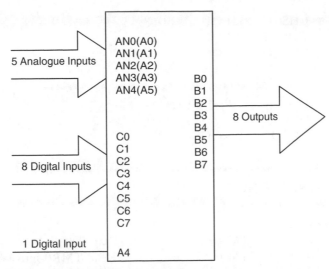

Figure 16.4 Port configuration of the 16F872

The 16F872 header

HEAD872.ASM
;EQUATES SECTION

TMR0	EQU	1
OPTION_R	EQU	1
PORTA	EQU	5
PORTB	EQU	6
PORTC	EQU	7
TRISA	EQU	5
TRISB	EQU	6
TRISC	EQU	7
STATUS	EQU	3
ZEROBIT	EQU	2
CARRY	EQU	0
EEADR	EQU	0DH
EEDATA	EQU	0CH
EECON1	EQU	0CH
EECON2	EQU	0DH
RD	EQU	0
WR	EQU	1
WREN	EQU	2
ADCON0	EQU	1FH
ADCON1	EQU	1FH
ADRES	EQU	1EH

```
              CHS0       EQU       3
              GODONE     EQU       2
              COUNT      EQU       20H


;*********************************************************
              LIST       P=16F872
              ORG        0
              GOTO       START


;*********************************************************
; SUBROUTINE SECTION.


;1 SECOND DELAY
DELAY1  CLRF    TMR0                    ;Start TMR0
LOOPA   MOVF    TMR0,W                  ;Read TMR0 into W
        SUBLW   .32                     ;TIME-W
        BTFSS   STATUS,ZEROBIT          ;Check TIME-W = 0
        GOTO    LOOPA
        RETLW   0                       ;Return after TMR0 = 32


;0.5 SECOND DELAY
DELAYP5 CLRF    TMR0                    ;Start TMR0
LOOPB   MOVF    TMR0,W                  ;Read TMR0 into W
        SUBLW   .16                     ;TIME-W
        BTFSS   STATUS,ZEROBIT          ;Check TIME-W = 0
        GOTO    LOOPB
        RETLW   0                       ;Return after TMR0 = 16


;*********************************************************************
;CONFIGURATION SECTION.


START       BSF        STATUS,5    ;Bank1
            MOVLW      B'11111111'
            MOVWF      TRISA       ;PortA is input

            MOVLW      B'00000000'
            MOVWF      TRISB       ;PortB is output

            MOVLW      B'11111111'
            MOVWF      TRISC       ;PortC is input

            MOVLW      B'00000111'
            MOVWF      OPTION_R    ;Option Register, TMR0 / 256
```

```
MOVLW    B'00000000'
MOVWF    ADCON1    ;PortA bits 0, 1, 2, 3, 5 are analogue
BSF      STATUS,6  ;BANK3
BCF      EECON1,7  ;Data memory on.
BCF      STATUS,5
BCF      STATUS,6  ;BANK0 return
BSF      ADCON0,0  ;turn on A/D.
CLRF     PORTA
CLRF     PORTB
CLRF     PORTC
```

.**
;
;Program starts now.

Explanation of HEAD872.ASM

Equates Section

- We have a third port, PORTC file 7 and its corresponding TRIS file, TRISC file 7 on Bank1. The TRIS file sets the I/O direction of the port bits.
- The EEPROM data file addresses have been included. EEADR is file 0Dh in Bank2, EEDATA is file 0Ch in Bank2, EECON is file 0Ch in Bank3 and EECON2 is file 0Dh in Bank3.
- The EEPROM data bits have been added. RD the read bit is bit 0, WR the write bit is bit 1, WREN the write enable bit is bit 2.
- The Analogue files ADRES, ADCON1 and ADCON2 have been included as have the associated bits CHS0 channel 0 select bit 3 and the GODONE bit, bit 2.

List Section

- This of course indicates the microcontroller being used, the 16F872 and that the first memory location is 0. In address 0 is the instruction GOTO START that instructs the micro to bypass the subroutine section and goto the configuration section at the label START.

Subroutine Section

- This includes the 2 delays DELAY1 and DELAYP5 as before.

Configuration Section

- As before we need to switch to Bank1 to address the TRIS files to configure the I/O. PORTA is set as an input port with the two instructions

```
MOVLW    B'00000111'
MOVWF    TRISA
```

PORTB and PORTC are configured in a similar manner using TRISB and TRISC.

- The Option register is configured with the instructions

```
MOVLW    B'00000111'
MOVWF    OPTION_R
```

- The A/D register is configured with the instructions

```
MOVLW    B'00000000'
MOVWF    ADCON1
```

Setting PORTA bits 0, 1, 2, 3 and 5 as analogue inputs.

- We turn to Bank3 by setting Bank select bit, STATUS,6 (bit 5 is still set) so that we can address EECON1, the EEPROM data control register. BSF EECON1 then enables access to the EEPROM program memory when required.
- We then turn back to Bank0 by clearing bits 5 and 6 of the Status register and clear the files PortA, PortB and PortC.

16F872 Application – a greenhouse control

In order to demonstrate the operation of the 16F872 and to develop our programming skills a little further consider the following application.

- A greenhouse has its temperature monitored so that a heater is turned on when the temperature drops below 15°C and turns the heater off when the temperature is above 17°C.
- A probe in the soil monitors the soil moisture so that a water valve will open for 5 seconds to irrigate the soil if it dries out. The valve is closed and will not be active for a further 5 seconds to give the water time to drain into the soil.
- A float switch monitors the level of the water and sounds an alarm if the water drops below a minimum level.

The circuit diagram for the greenhouse control is shown in Figure 16.5 and the flowchart is drawn in Figure 16.6.

Greenhouse program

In order to program the analogue/digital settings consider the NTC Thermister. As the temperature increases the resistance of the thermister will decrease and so the voltage presented to AN0 will increase.

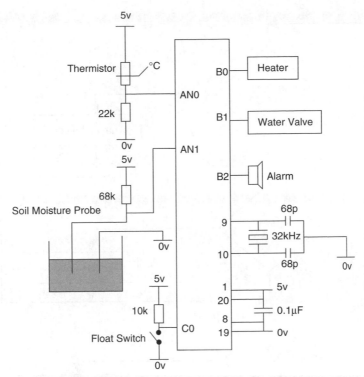

Figure 16.5 Greenhouse control circuit

Let us assume the voltage is 2.9v at 15°C and 3.2v at 17°C they correspond to digital readings of $2.9 \times 51 = 147.9$ i.e. 148 and $3.2 \times 51 = 163.2$ i.e. 163. (N.B. 5v = 255, so 1v = 51 we are using an 8 bit A/D.)

Our program then needs to check when AN0 goes above 163 and below 148.

As the soil dries out its resistance will increase. Let us assume in our application dry soil will give a reading of 2.6v, (on AN1), i.e. $2.6 \times 51 = 132.6$ i.e. 133. So any reading above 133 is considered dry.

The float switch is a digital input showing 1 if the water level is above the minimum required and a 0 if it is below the minimum.

Greenhouse code

The code for the greenhouse uses HEAD872.ASM with the program instuctions added and saved as GREENHO.ASM.

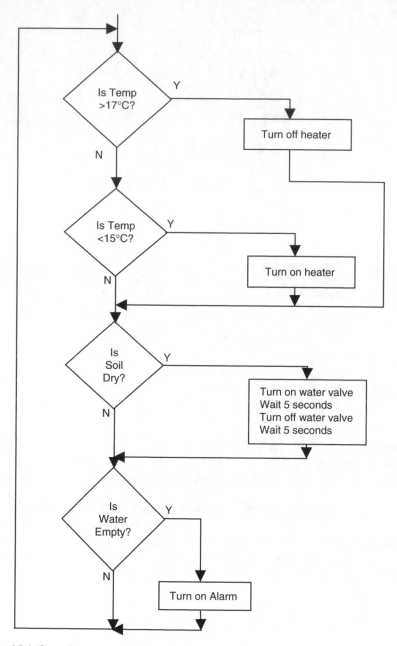

Figure 16.6 Greenhouse control flowchart

```
;GREENHO.ASM
;EQUATES SECTION
                TMR0        EQU     1
                OPTION_R    EQU     1
                PORTA       EQU     5
                PORTB       EQU     6
                PORTC       EQU     7
                TRISA       EQU     5
                TRISB       EQU     6
                TRISC       EQU     7
                STATUS      EQU     3
                ZEROBIT     EQU     2
                CARRY       EQU     0
                EEADR       EQU     0DH
                EEDATA      EQU     0CH
                EECON1      EQU     0CH
                EECON2      EQU     0DH
                RD          EQU     0
                WR          EQU     1
                WREN        EQU     2
                ADCON0      EQU     1FH
                ADCON1      EQU     1FH
                ADRES       EQU     1EH
                CHS0        EQU     3
                GODONE      EQU     2
                COUNT       EQU     20H
```

```
;****************************************************
                LIST        P=16F872
                ORG         0
                GOTO        START
;****************************************************
;Configuration Bits

__CONFIG H'3F30'          ;selects LP oscillator, WDT off, PUT on,
                          ;Code Protection disabled.
;****************************************************
;SUBROUTINE SECTION.

;1 SECOND DELAY
DELAY1   CLRF    TMR0                    ;Start TMR0
LOOPA    MOVF    TMR0,W                  ;Read TMR0 into W
         SUBLW   .32                     ;TIME-W
         BTFSS   STATUS,ZEROBIT  ;Check TIME-W = 0
```

```
                GOTO      LOOPA
                RETLW     0                        ;Return after TMR0 = 32 >

;0.5 SECOND DELAY
DELAYP5  CLRF     TMR0                    ;Start TMR0
LOOPB    MOVF     TMR0,W                  ;Read TMR0 into W
         SUBLW    .16                     ;TIME-W
         BTFSS    STATUS,ZEROBIT          ;Check TIME-W = 0
         GOTO     LOOPB
         RETLW    0                       ;Return after TMR0 = 16

;5 SECOND DELAY
DELAY5   CLRF     TMR0                    ;Start TMR0
LOOPC    MOVF     TMR0,W                  ;Read TMR0 into W
         SUBLW    .160                    ;TIME-W
         BTFSS    STATUS,ZEROBIT          ;Check TIME-W = 0
         GOTO     LOOPC
         RETLW    0                       ;Return after TMR0 = 160

HEAT_ON    BSF        PORTB,0    ;Turn heater on
           GOTO       SOIL       ;Check soil moisture

HEAT_OFF   BCF        PORTB,0    ;Turn heater off
           GOTO       SOIL       ;Check soil moisture

WATER_ON BSF          PORTB,1    ;Turn water on
           CALL       DELAY5
           BCF        PORTB,1    ;Turn water off
           CALL       DELAY5
           GOTO       WATER      ;Check water level

ALARM_ON BSF          PORTB,2    ;Turn alarm on
           GOTO       BEGIN      ;Repeat the process

ALARM_OFF BCF         PORTB,2    ;Turn alarm off
           GOTO       BEGIN      ;Repeat the process
;****************************************************************
; CONFIGURATION SECTION.

START      BSF        STATUS,5   ;Bank1
           MOVLW      B'11111111'
           MOVWF      TRISA      ;PortA is input

           MOVLW      B'00000000'
           MOVWF      TRISB      ;PortB is output
```

```
           MOVLW    B'11111111'
           MOVWF    TRISC        ;PortC is input

           MOVLW    B'00000111'
           MOVWF    OPTION_R     ;Option Register, TMR0/256

           MOVLW    B'00000000'
           MOVWF    ADCON1       ;PortA bits 0, 1, 2, 3, 5 are
                                 ;analogue
           BSF      STATUS,6     ;BANK3
           BCF      EECON1,7     ;Data memory on.
           BCF      STATUS,5
           BCF      STATUS,6     ;BANK0 return
           BSF      ADCON0,0     ;turn on A/D.
           CLRF     PORTA
           CLRF     PORTB
           CLRF     PORTC
```

;**
;Program starts now.

;Check the temperature on AN0
```
BEGIN      BCF      ADCON0,CHS0      ;C to select AN0
           BSF      ADCON0,GODONE
WAIT1      BTFSC    ADCON0,GODONE
           GOTO     WAIT1
           MOVF     ADRES,W
           SUBLW    .163             ;163 – W

           BTFSS    STATUS,CARRY     ;C if W > 163 i.e. hot
                                     ;(above 17°C)

           GOTO     HEAT_OFF

           MOVF     ADRES,W
           SUBLW    .148             ;148 – W

           BTFSC    STATUS,CARRY     ;S if W < 148 i.e. cold
                                     ;(below 15°C)

           GOTO     HEAT_ON
```

;Check the soil moisture on AN1
```
SOIL       BSF      ADCON0,CHS0      ;S to select AN1
           BSF      ADCON0,GODONE
WAIT2      BTFSC    ADCON0,GODONE
           GOTO     WAIT2
```

```
            MOVF      ADRES,W
            SUBLW     .133                ;133 – W
            BTFSS     STATUS,CARRY        ;C if W > 133 i.e. dry
            GOTO      WATER_ON

;Check water is above minimum
WATER       BTFSC     PORTC,0             ;C if below minimum
            GOTO      ALARM_OFF
            GOTO      ALARM_ON
END
```

Explanation of code

In the previous analogue circuits in Chapter 11 we only used 1 analogue input on AN0. We now have two analogue inputs on AN0 and AN1. When making an analogue measurement we must specify which analogue channel we wish to measure. The default is AN0 when moving to AN1 we select AN1 by setting channel select bit0 i.e. BSF ADCON0,CHS0.

When moving back to AN0 clear the channel select bit. The 8 channels, AN0 to AN7 are seclected using bits, CHS2, CHS1, CHS0.

- The temperature is read on AN0 with and then checked to see if it is greater than 17°C, by subtracting the A/D reading from 163 (the reading equating to 17°C). The carry bit in the status register indicates if the result is +ve or −ve being set or clear. We then go to turn off the heater if the temperature is above 17°C or check if the temperature is below 15°C. In which case we turn on the heater.
- The soil moisture is checked next. AN1 is selected and the reading compared this time to 133 indicating dry soil. The program either goes to turn on the water valve if the soil is dry or continues to check the water level if the soil is wet.
- If the water level is below minimum then the alarm sounds, if above minimum the alarm is turned off. The program then repeats the checking of the inputs and reacts to them accordingly.

Programming the 16F872 microcontroller using PICSTART PLUS

Once the pogram GREENHO.ASM has been saved it is then assembled using MPASMWIN. The next step as previously is to program GREENHO.HEX into the micro using PICSTART PLUS.

This process has been outlined in Chapter 2, but there are a few more selections to attend to in the 'Device Specification' Section.

- Select the device 16F872, if this device is not available you will require a later version of MPLAB, obtainable from www.microchip.com.
- Set the fuses.

Configuration bits

The configuration bit settings when programming the 16F872 for the Greenhouse program are shown in Figure 16.7.

Address	Value	Category	Setting
2007	3F30	Oscillator	LP
		Watchdog Timer	Off
		Power Up Timer	On
		Brown Out Detect	Off
		Low Voltage Program	Disabled
		Flash Program Write	Enabled
		Background Debug	Disabled
		Data EE Read Protect	Off
		Code Protect	Off

Figure 16.7 Greenhouse program configuration bits

Reconfiguring the 16F872 header

- The port settings are changed as they were for the 16F84 i.e. a 1 means the pin is an input and a 0 means an output.
- The Option Register is configured as in the 16F84 see also Chapter 19.
- The A/D convertor configuration is adjusted using A/D configuration register 1, i.e. ADCON1 shown in Figure 16.8.

ADFM				PCFG3	PCFG2	PCFG1	PCFG0
bit7							bit0

Figure 16.8 ADCON1, A/D port configuration register 1

Bit7 is the A/D Format Select bit, which selects which bits of the A/D result registers are used. I.e. the A/D can use 10 bits which requires two result registers, ADRESH and ADRESL. Two formats are available.

(a) the most significant bits of ADRESH read as 0, with $ADFM = 1$

		ADRERSH					
0	0	0	0	0	0		

			ADRESL				

Or (b) the least significant bits of ADRESL read as 0, with $ADFM = 0$

ADRESH

ADRESL

		0	0	0	0	0	0

For 8 bit operation condition (b) is used with ADRESH as the 8 most significant bits of the A/D result. This is the default configuration used in HEADER872.ASM where ADRESH (ADRES in the equates) is register 1Eh in Bank0.

Table 16.1 A/D Port configuration

PCFG3: PCFG0	AN7 E2	AN6 E1	AN5 E0	AN4 A5	AN3 A3	AN2 A2	AN1 A1	AN0 A0	Vref+	Vref−
0000	A	A	A	A	A	A	A	A	Vdd	Vss
0001	A	A	A	A	Vref+	A	A	A	A3	Vss
0010	D	D	D	A	A	A	A	A	Vdd	Vss
0011	D	D	D	A	Vref+	A	A	A	A3	Vss
0100	D	D	D	D	A	D	A	A	Vdd	Vss
0101	D	D	D	D	Vref+	D	A	A	A3	Vss
011X	D	D	D	D	D	D	D	D	Vdd	Vss
1000	A	A	A	A	Vref+	Vref−	A	A	A3	A2
1001	D	D	A	A	A	A	A	A	Vdd	Vss
1010	D	D	A	A	Vref+	A	A	A	A3	Vss
1011	D	D	A	A	Vref+	Vref−	A	A	A3	A2
1100	D	D	D	A	Vref+	Vref−	A	A	A3	A2
1101	D	D	D	D	Vref+	Vref−	A	A	A3	A2
1110	D	D	D	D	D	D	D	A	Vdd	Vss
1111	D	D	D	D	Vref+	Vref−	D	A	A3	A2

Table 16.1 shows the A/D Port Configuration settings for PCFG3, PCFG2, PCFG1 and PCFG0.

A = Analogue Input, D = Digital input.
Vdd = +ve supply, Vss = −ve supply.
Vref+ = high voltage reference.
Vref− = low voltage reference.
A3 = PortA,3 A2 = PortA,2 etc.

N.B. AN7, AN6 and AN5 are only available on the 40 pin devices 16F871, 16F874 and 16F877.

17
The 16F62X microcontroller

The 16F62X family of microcontrollers includes the two devices 16F627 and 16F628.

The 16F62X microcontrollers are flash devices and have 18 pins and data EEPROM just like the 16F84, but they have more functions. Notably there is an on board oscillator so an external crystal is not required. This frees up two pins for extra I/O. The 16F62X in fact can use 16 of its 18 pins as I/O.

Table 17.1 shows the specification of the 16F62X devices and the 16F84 for comparison.

Table 17.1 The 16F62X specification

Device	Flash Program Memory (bytes)	RAM Data Memory (bytes)	EEPROM Data Memory (bytes)	Timer Modules	I/O Pins
16F627	1024	224	128	3	16
16F628	2048	224	128	3	16
16F84	1024	68	64	1	13

16F62X oscillator modes

The 16F62X can be operated in 8 different oscillator modes. They are selected when programming the device just like the 16F84, or by inserting the configuration bits in the header.

The options are:

- LP   Low Power Crystal, 32.768kHz
- XT   4MHz Crystal
- HS   High Speed Crystal, 20MHz
- ER   External Resistor (2 modes)
- INTRC   Internal Resistor/Capacitor (2 modes)
- EC   External Clock in

The two modes for the internal resistor/capacitor configuration are 4MHz and 37kHz. The default setting is 4MHz. The 16F627 header, HEAD62RC.ASM, selects the 37kHz oscillator by clearing the OSCF (oscillator frequency) bit, bit3 in the Peripheral Control Register, PCON with BCF PCON,3.

There was obviously a good reason for Microchip choosing 37kHz for the oscillator instead of 32.768kHz, I only wish I knew what it was! 32.768kHz as we have seen before (HEADER84.ASM) can give us TMR0 pulses of 32 a second when setting the option register to divide the program timing pulses by 256.

The most attractive proposition I can see using 37kHz is:

- Clock frequency = 37kHz,
- Program execution frequency is 37kHz/4 = 9250Hz.
- Setting the prescaler to /32 gives TMR0 pulses of 9250 / 32 = 289.0625Hz = 0.03459459s for each pulse.
- Counting 29 TMR0 pulses gives a time of 0.100324324s i.e. 0.1s + 0.3% error. If this error, about 4.5 minutes a day, is unacceptable then a 32.768kHz crystal can be used as we did with the 16F84.

Since the programs used previously on the 16F84 did not require any accurate timing our 16F62X header will set the prescaler to divide by 32 and use a subroutine to count 29 TMR0 pulses to give a time of 0.1s.

All of the 16F84 programs can then be transferred to the 16F62X header.

The choice of a 32.768kHz crystal or the 37kHz internal RC will obviously make a difference to the timing routines in the header. I have therefore included two headers for the 16F62X devices. HEAD62LP.ASM for use with the 32kHz crystal and HEAD62RC.ASM for use with the 37kHz internal RC oscillator.

16F62X and 16F84 Pinouts

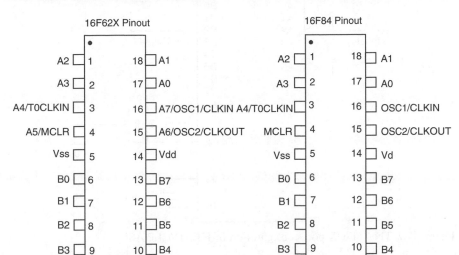

16F62X Port configuration

The header (HEAD62RC.ASM) will configure the 16F62X I/O as shown in Figure 17.1.

The header (HEAD62LP.ASM) will configure the 16F62X I/O as shown in Figure 17.2.

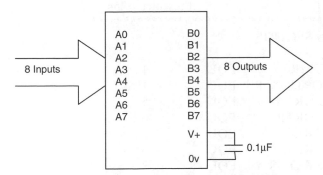

Figure 17.1 The 16F62X port configuration in HEAD62RC.ASM

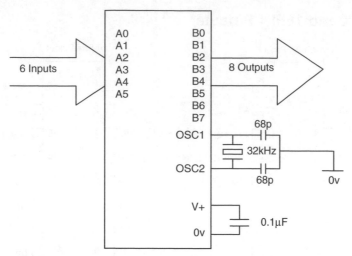

Figure 17.2 The 16F62X port configuration in HEAD62LP.ASM

16F62X Memory map

The 16F62X Memory Map at the end of the chapter (page 256).

The 16F62X headers

HEAD62LP.ASM

```
;HEAD62LP.ASM using the 32kHz crystal
                        ;PortA bits 0 to 5 are inputs
                        ;PortB bits 0 to 7 are outputs
                        ;Prescaler / 256
;*********************************************
;EQUATES SECTION

        TMR0        EQU        1
        OPTION_R    EQU        1
        PORTA       EQU        5
        PORTB       EQU        6
        TRISA       EQU        5
        TRISB       EQU        6
        STATUS      EQU        3
        ZEROBIT     EQU        2
        CARRY       EQU        0
        EEADR       EQU        1BH
        EEDATA      EQU        1AH
```

```
            EECON1    EQU       1CH
            EECON2    EQU       1DH
            RD        EQU       0
            WR        EQU       1
            WREN      EQU       2
            COUNT     EQU       20H
```

;**
            LIST   P=16F627   ;using the 627
            ORG   0
            GOTO   START

;**
Configuration Bits

__CONFIG H'3F00'   ;selects LP oscillator, WDT off,
                    ;Code Protection disabled.

;**
;SUBROUTINE SECTION.

;1 SECOND DELAY
```
DELAY1  CLRF    TMR0                      ;Start TMR0
LOOPA   MOVF    TMR0,W                    ;Read TMR0 into W
        SUBLW   .32                       ;TIME-W
        BTFSS   STATUS,ZEROBIT ;Check TIME-W=0
        GOTO    LOOPA
        RETLW   0                         ;Return after TMR0 = 32
```

;0.5 SECOND DELAY
```
DELAYP5 CLRF    TMR0                      ;Start TMR0
LOOPB   MOVF    TMR0,W                    ;Read TMR0 into W
        SUBLW   .16                       ;TIME-W
        BTFSS   STATUS,ZEROBIT ;Check TIME-W=0
        GOTO    LOOPB
        RETLW   0                         ;Return after TMR0 = 16
```

;**
;CONFIGURATION SECTION.

```
START   BSF     STATUS,5    ;Bank1
        MOVLW   B'11111111'
        MOVWF   TRISA       ;PortA is input
```

```
              MOVLW       B'00000000'
              MOVWF       TRISB        ;PortB is output

              MOVLW       B'00000111'
              MOVWF       OPTION_R     ;Option Register, TMR0/256

              BCF         STATUS,5     ;Bank0
              CLRF        PORTA
              CLRF        PORTB

              MOVLW       .7
              MOVWF       1FH          ;CMCON turns off comparators.
```

;***
;Program starts now.

END

HEAD62RC.ASM

;HEAD62RC.ASM using the 37kHz internal RC
                            ;PortA bits 0 to 7 are inputs
                            ;PortB bits 0 to 7 are outputs
                            ;Prescaler/32

;***
;EQUATES SECTION

```
              TMR0        EQU          1
              OPTION_R    EQU          1
              PORTA       EQU          5
              PORTB       EQU          6
              TRISA       EQU          5
              TRISB       EQU          6
              STATUS      EQU          3
              ZEROBIT     EQU          2
              CARRY       EQU          0
              EEADR       EQU          1BH
              EEDATA      EQU          1AH
              EECON1      EQU          1CH
              EECON2      EQU          1DH
              RD          EQU          0
              WR          EQU          1
              WREN        EQU          2
              PCON        EQU          0EH
              COUNT       EQU          20H
```

```
;************************************************
            LIST      P=16F627    ;using the 627
            ORG       0
            GOTO      START

;************************************************
Configuration Bits

__CONFIG H'3F10'      ;selects Internal RC oscillator, WDT off,
                      ;Code Protection disabled.

;************************************************
;SUBROUTINE SECTION.

;0.1 SECOND DELAY
DELAYP1  CLRF    TMR0                ;Start TMR0
LOOPA    MOVF    TMR0,W              ;Read TMR0 into W
         SUBLW   .29                 ;TIME-W
         BTFSS   STATUS,ZEROBIT      ;Check TIME-W=0
         GOTO    LOOPA
         RETLW   0                   ;Return after TMR0 = 29

;0.5 SECOND DELAY
DELAYP5  MOVLW   5
         MOVWF   COUNT
LOOPB    CALL    DELAYP1    ;0.1s delay
         DECFSZ  COUNT
         GOTO    LOOPB
         RETLW   0                   ;Return after 5 DELAYP1

;1 SECOND DELAY
DELAY1   MOVLW   10
         MOVWF   COUNT
LOOPC    CALL    DELAYP1    ;0.1s delay
         DECFSZ  COUNT
         GOTO    LOOPC
         RETLW   0                   ;Return after 10 DELAYP1

;************************************************
;CONFIGURATION SECTION.

START    BSF     STATUS,5   ;Bank1
         MOVLW   B'11111111'
         MOVWF   TRISA      ;PortA is input
```

```
          MOVLW     B'00000000'
          MOVWF     TRISB       ;PortB is output

          MOVLW     B'00000100'
          MOVWF     OPTION_R   ;Option Register, TMR0 / 32
          CLRF      PCON        ;Select 37kHz oscillator.
          BCF       STATUS,5    ;Bank0
          CLRF      PORTA
          CLRF      PORTB

          MOVLW     .7
          MOVWF     1FH         ;CMCON turns off comparators.
```

;***
;Program starts now.

A 16F627 application – flashing an LED on and off

In order to introduce the operation of the 16F672 device we will consider the simple example of the single LED flashing on and off, which was introduced in Chapter 2.

The 16F627 will be operated in the INTRC mode using the internal 37kHz oscillator.

The circuit diagram for this is shown in Figure 17.3.

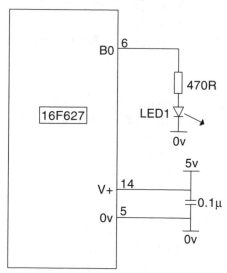

Figure 17.3 The 16F627 LED flashing circuit

The 16F627 LED flasher code

;FLASH_RC.ASM using the 37kHz internal RC
;PortA bits 0 to 7 are inputs
;PortB bits 0 to 7 are outputs
;Prescaler/32

```
;*********************************************
;EQUATES SECTION

           TMR0      EQU      1
           OPTION_R  EQU      1
           PORTA     EQU      5
           PORTB     EQU      6
           TRISA     EQU      5
           TRISB     EQU      6
           STATUS    EQU      3
           ZEROBIT   EQU      2
           CARRY     EQU      0
           EEADR     EQU      1BH
           EEDATA    EQU      1AH
           EECON1    EQU      1CH
           EECON2    EQU      1DH
           RD        EQU      0
           WR        EQU      1
           WREN      EQU      2
           PCON      EQU      0EH
           COUNT     EQU      20H

;****************************************************
           LIST      P=16F627    ;using the 627
           ORG       0
           GOTO      START

;*******************************************************
;Configuration Bits

__CONFIG H'3F10'      ;selects Internal RC oscillator, WDT off,
                      ;Code Protection disabled.

;*******************************************************
;SUBROUTINE SECTION.

;0.1 SECOND DELAY
DELAYP1   CLRF       TMR0        ;Start TMR0
```

```
LOOPA      MOVF      TMR0,W              ;Read TMR0 into W
           SUBLW     .29                 ;TIME-W
           BTFSS     STATUS,ZEROBIT      ;Check TIME-W=0
           GOTO      LOOPA
           RETLW     0                   ;Return after TMR0 = 29


;0.5 SECOND DELAY
DELAYP5    MOVLW     5
           MOVWF     COUNT
LOOPB      CALL      DELAYP1             ;0.1s delay
           DECFSZ    COUNT
           GOTO      LOOPB
           RETLW     0                   ;Return after 5 DELAYP1


;****************************************************************
;CONFIGURATION SECTION.

START      BSF       STATUS,5            ;Bank1
           MOVLW     B'11111111'
           MOVWF     TRISA               ;PortA is input

           MOVLW     B'00000000'
           MOVWF     TRISB               ;PortB is output

           MOVLW     B'00000100'
           MOVWF     OPTION_R            ;Option Register, TMR0 / 32
           CLRF      PCON                ;Selects 37kHz oscillator.
           BCF       STATUS,5            ;Bank0
           CLRF      PORTA
           CLRF      PORTB

           MOVLW     .7
           MOVWF     1FH                 ;CMCON turns off comparators.

;********************************************************
;Program starts now.
BEGIN      BSF       PORTB,0             ;Turn on LED
           CALL      DELAYP5             ;Wait 0.5s
           BCF       PORTB,0             ;Turn off LED
           CALL      DELAYP5             ;Wait 0.5s
           GOTO      BEGIN               ;Repeat
END
```

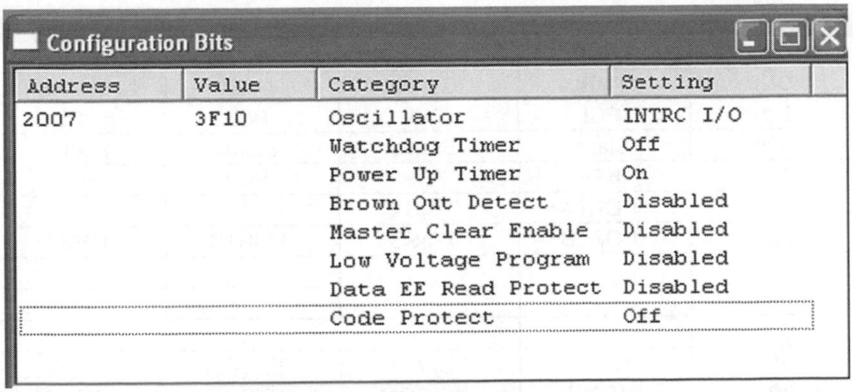

Address	Value	Category	Setting
2007	3F10	Oscillator	INTRC I/O
		Watchdog Timer	Off
		Power Up Timer	On
		Brown Out Detect	Disabled
		Master Clear Enable	Disabled
		Low Voltage Program	Disabled
		Data EE Read Protect	Disabled
		Code Protect	Off

Figure 17.4 Configuration settings for FLASH_RC.HEX

The operation of the program after 'Program starts now', is exactly the same as in FLASHER.ASM in Chapter 2, using the 16F84.

All of the programs using the 16F84 can be transferred by copying the code starting at 'Program starts now' and pasting into HEAD62RC.ASM or HEAD62LP.ASM as required.

Configuration settings for the 16F627

When programming the Code FLASH_RC.HEX into the 16F627 use the configuration settings shown in Figure 17.4. This setting equates to H'3F10' which can be written into the Configuration Bits setting in your code.

Other features of the 16F62X

The 16F62X also includes,

- An analogue comparator module with 2 analogue comparators and an on-chip voltage reference module.
- Timer1 a 16 bit timer/counter module with external crystal/clock capability and Timer2 an 8 bit timer/counter with prescaler and postscaler.
- A Capture, Compare and Pulse Width Modulation modes.

Please refer to the 16F62X data sheet for operation of these other features.

Address	File Name	File Name	File Name	File Name
00h	Ind.Add	Ind.Add	Ind.Add	Ind.Add
01h	TMR0	Option	TMR0	Option
02h	PCL	PCL	PCL	PCL
03h	Status	Status	Status	Status
04h	FSR	FSR	FSR	FSR
05h	PORTA	TRISA		
06h	PORTB	TRISB	PORTB	TRISB
07h				
08h				
09h				
0Ah	PCLATH	PCLATH	PCLATH	PCLATH
0Bh	INTCON	INTCON	INTCON	INTCON
0Ch	PIR1	PIE1		
0Dh				
0Eh	TMR1L	PCON		
0Fh	TMR1H			
10h	T1CON			
11h	TMR2			
12h	T2CON	PR2		
13h				
14h				
15h	CCPR1L			
16h	CCPR1H			
17h	CCP1CON			
18h	RCSTA	TXSTA		
19h	TXREG	SPBRG		
1Ah	RCREG	EEDATA		
1Bh		EEADR		
1Ch		EECON1		
1Dh		EECON2		
1Eh				
1Fh	CMCON	VRCON		
.	General Purpose Register 96 bytes	General Purpose Register 80 bytes	General Purpose Register 48 bytes	
6Fh				
7F h				
	Bank0	Bank1	Bank2	Bank3

The 16F62X memory map

18
Projects

Project 1   Electronic dice

When using a Microcontroller in a control system the place to start is to decide what hardware you are controlling. In the Electronic Dice we will use 7 LEDs for the display and a push button to make the "throw". Just to make the dice a little more interesting we will use a buzzer to give an audible indication of the number thrown.

The circuit for the Dice is shown in Figure 18.1, using the 16F818 with its internal 31.25kHz clock. The push button is an input connected to PortA,2. The 7 LEDs are connected to PortB and the buzzer is on A1.

The truth table for the dice is shown in Table 18.1.

How does it work?

The dice has an input – the "throw" button. When it is pressed the internal count repeatedly runs through from 1 to 6 changing some 8000 times a second and stops on a number when the button is released.

This would be a complicated circuit to design with a timer, counter and decoder circuits. But now we can use one chip to do all the timing counting and decoding functions. Not only that I have also added a light flashing routine for the first few seconds when the dice is turned on. Try doing all that with one chip – other than a microcontroller.

The best way to describe the action of a program is with a flowchart. The flowchart for the dice is shown in Figure 18.2.

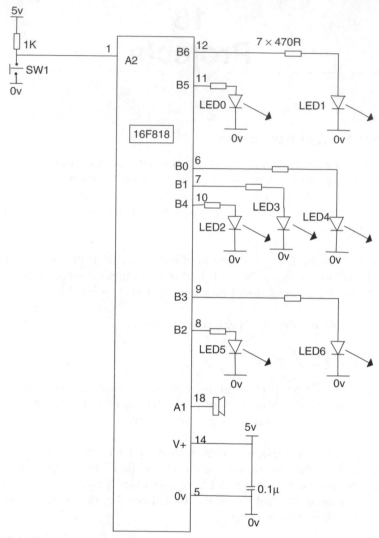

Figure 18.1 Circuit diagram for the electronic dice

Table 18.1 Truth table for the electronic dice

Throw	B7	B6	B5	B4	B3	B2	B1	B0
1	0	0	0	0	0	0	1	0
2	0	0	1	0	1	0	0	0
3	0	0	1	0	1	0	1	0
4	0	1	1	0	1	1	0	0
5	0	1	1	0	1	1	1	0
6	0	1	1	1	1	1	0	1

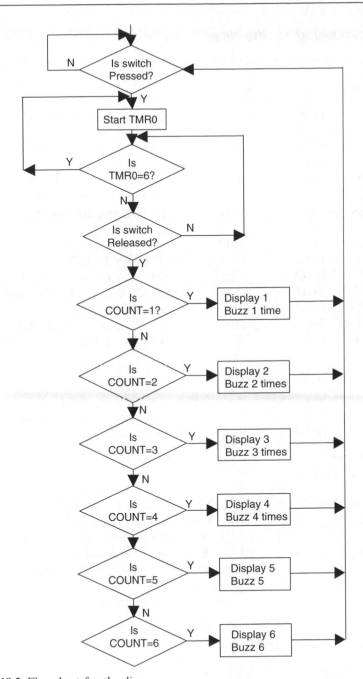

Figure 18.2 Flowchart for the dice

Program listing for the dice

The full program listing for the dice is given below in ;DICE.ASM.

;DICE.ASM

TMR0	EQU	1	;means TMR0 is file 1.
PC	EQU	2	
STATUS	EQU	3	;means STATUS is file 3.
PORTA	EQU	5	;means PORTA is file 5.
PORTB	EQU	6	;means PORTB is file 6.
ZEROBIT	EQU	2	;means ZEROBIT is bit 2.
ADCON0	EQU	1FH	;A/D Configuration reg.0
ADCON1	EQU	9FH	;A/D Configuration reg.1
ADRES	EQU	1EH	;A/D Result register.
CARRY	EQU	0	;CARRY IS BIT 0.
TRISA	EQU	85H	;PORTA Configuration Register
TRISB	EQU	86H	;PORTB Configuration Register
OPTION_R	EQU	81H	;Option Register
OSCCON	EQU	8FH	;Oscillator control register.
COUNT	EQU	20H	;COUNT a register to count events.
COUNTA	EQU	21H	

;**

	LIST	P=16F818	;we are using the 16F818.
	ORG	0	;the start address in memory is 0
	GOTO	START	;goto start!

;**
;Configuration Bits

__CONFIG H'3F10'   ;sets INTRC-A6 is port I/O, WDT off, PUT on,
                        ;MCLR tied to VDD A5 is I/O
                        ;BOD off, LVP disabled, EE protect disabled,
                        ;Flash Program Write disabled,
                        ;Background Debugger Mode disabled, CCP
                        ;function on B2,
                        ;Code Protection disabled.

;**
;SUBROUTINE SECTION.

;0.1 second delay, actually 0.099968s

DELAYP1	CLRF	TMR0	;START TMR0.
LOOPB	MOVF	TMR0,W	;READ TMR0 INTO W.

```
         SUBLW     .3              ;TIME-3
         BTFSS     STATUS,
                   ZEROBIT         ;Check TIME-W = 0
         GOTO      LOOPB           ;Time is not = 3.
         NOP                       ;add extra delay
         NOP
         RETLW     0               ;Time is 3, return.
```

;0.3 second delay.

```
DELAY    MOVLW     .3
         MOVWF     COUNT
LOOPC    CALL      DELAYP1
         DECFSZ    COUNT
         GOTO      LOOPC
         RETLW     0
```

;1 second delay.

```
DELAY1   MOVLW     .10
         MOVWF     COUNT
LOOPA    CALL      DELAYP1
         DECFSZ    COUNT
         GOTO      LOOPA
         RETLW     0
```

;**
;CONFIGURATION SECTION.

```
START    BSF       STATUS,5     ;Turns to Bank1.

         MOVLW     B'11111101'  ;7 bits of PORTA are I/P
         MOVWF     TRISA

         MOVLW     B'00000110'  ;PORTA IS DIGITAL
         MOVWF     ADCON1

         MOVLW     B'00000000'
         MOVWF     TRISB        ;PORTB is OUTPUT

         MOVLW     B'00000000'
         MOVWF     OSCCON       ;oscillator 31.25kHz

         MOVLW     B'00000111'  ;Prescaler is /256
         MOVWF     OPTION_R     ;TIMER is 1/32 secs.
```

```
            BCF         STATUS,5     ;Return to Bank0.
            CLRF        PORTA        ;Clears PortA.
            CLRF        PORTB        ;Clears PortB.


;***********************************************************
;Program starts now.
            CALL        DELAY1
            CALL        DELAY1
            CLRF        PORTB        ;Turn off LEDs and buzzer.
            MOVLW       .5
            MOVWF       COUNTA

SEC1        MOVLW       60H          ;Light flashing routine.
            MOVWF       PORTB
            CALL        DELAY
            MOVLW       13H
            MOVWF       PORTB
            CALL        DELAY
            MOVLW       0CH
            MOVWF       PORTB
            CALL        DELAY
            MOVLW       13H
            MOVWF       PORTB
            CALL        DELAY
            DECFSZ      COUNTA
            GOTO        SEC1

            CALL        DELAY1
            BSF         PORTA,1      ;Turn buzzer on
            CALL        DELAY1
            BCF         PORTA,1      ;Turn buzzer off

BEGIN       BTFSC       PORTA,2      ;Is switch pressed?
            GOTO        BEGIN        ;NO
            CALL        DELAYP1      ;YES
            CLRF        PORTB        ;Switch off LEDs
LOOP1       CLRF        TMR0         ;Start Timer
LOOP2       MOVF        TMR0,W       ;Put time into W.
            SUBLW       6            ;Is TMR0 = 6?
            BTFSC       STATUS,
                        ZEROBIT      ;Skip if TMR0 is not 6.
            GOTO        LOOP1        ;TMR0 is 6, so reset timer.
            BTFSS       PORTA,2      ;skip if button released?
            GOTO        LOOP2        ;No, Carry on timing
```

```
              MOVF       TMR0,W      ;yes, put the TMR0 into W.
              ADDWF      PC          ;Jump the value of W.
              GOTO       NUM1        ;TMR0=0
              GOTO       NUM2        ;TMR0=1
              GOTO       NUM3        ;TMR0=2
              GOTO       NUM4        ;TMR0=3
              GOTO       NUM5        ;TMR0=4
              GOTO       NUM6        ;TMR0=5

NUM1          MOVLW      B'00000010' ;Turn LED on
              MOVWF      PORTB
              BSF        PORTA,1     ;turn on buzzer for 1/4 sec.
              CALL       DELAY
              BCF        PORTA,1     ;Turn buzzer off.
              GOTO       BEGIN       ;BEGIN AGAIN.

NUM2          MOVLW      B'00101000' ;TURN ON 2 LEDS.
              MOVWF      PORTB
              BSF        PORTA,1     ;turn on buzzer for 1/4 sec.
              CALL       DELAY
              BCF        PORTA,1     ;turn off buzzer for 1/4 sec.
              CALL       DELAY
              BSF        PORTA,1     ;turn on buzzer for 1/4 sec.
              CALL       DELAY
              BCF        PORTA,1     ;Turn buzzer off.
              GOTO       BEGIN

NUM3          MOVLW      B'00101010'
              MOVWF      PORTB
              BSF        PORTA,1     ;turn on buzzer for 1/4 sec.
              CALL       DELAY
              BCF        PORTA,1     ;turn off buzzer for 1/4 sec.
              CALL       DELAY
              BSF        PORTA,1     ;turn on buzzer for 1/4 sec.
              CALL       DELAY
              BCF        PORTA,1     ;turn off buzzer for 1/4 sec.
              CALL       DELAY
              BSF        PORTA,1     ;turn on buzzer for 1/4 sec.
              CALL       DELAY
              BCF        PORTA,1     ;Turn off buzzer.
              GOTO       BEGIN

NUM4          MOVLW      B'01101100'
              MOVWF      PORTB
```

```
              BSF      PORTA,1    ;turn on buzzer for 1/4 sec.
              CALL     DELAY
              BCF      PORTA,1    ;turn off buzzer for 1/4 sec.
              CALL     DELAY
              BSF      PORTA,1    ;turn on buzzer for 1/4 sec.
              CALL     DELAY
              BCF      PORTA,1    ;turn off buzzer for 1/4 sec.
              CALL     DELAY
              BSF      PORTA,1    ;turn on buzzer for 1/4 sec.
              CALL     DELAY
              BCF      PORTA,1    ;turn off buzzer for 1/4 sec.
              CALL     DELAY
              BSF      PORTA,1    ;turn on buzzer for 1/4 sec.
              CALL     DELAY
              BCF      PORTA,1    ;Turn buzzer off.

              GOTO     BEGIN

NUM5          MOVLW    B'01101110'
              MOVWF    PORTB
              BSF      PORTA,1    ;turn on buzzer for 1/4 sec.
              CALL     DELAY
              BCF      PORTA,1    ;turn off buzzer for 1/4 sec.
              CALL     DELAY
              BSF      PORTA,1    ;turn on buzzer for 1/4 sec.
              CALL     DELAY
              BCF      PORTA,1    ;turn off buzzer for 1/4 sec.
              CALL     DELAY
              BSF      PORTA,1    ;turn on buzzer for 1/4 sec.
              CALL     DELAY
              BCF      PORTA,1    ;turn off buzzer for 1/4 sec.
              CALL     DELAY
              BSF      PORTA,1    ;turn on buzzer for 1/4 sec.
              CALL     DELAY
              BCF      PORTA,1    ;turn off buzzer for 1/4 sec.
              CALL     DELAY
              BSF      PORTA,1    ;turn on buzzer for 1/4 sec.
              CALL     DELAY
              BCF      PORTA,1    ;turn off buzzer.
              GOTO     BEGIN

NUM6          MOVLW    B'01111101'
              MOVWF    PORTB
              BSF      PORTA,1    ;turn on buzzer for 1/4 sec.
              CALL     DELAY
```

```
            BCF      PORTA,1    ;turn off buzzer for 1/4 sec.
            CALL     DELAY
            BSF      PORTA,1    ;turn on buzzer for 1/4 sec.
            CALL     DELAY
            BCF      PORTA,1    ;turn off buzzer for 1/4 sec.
            CALL     DELAY
            BSF      PORTA,1    ;turn on buzzer for 1/4 sec.
            CALL     DELAY
            BCF      PORTA,1    ;turn off buzzer for 1/4 sec.
            CALL     DELAY
            BSF      PORTA,1    ;turn on buzzer for 1/4 sec.
            CALL     DELAY
            BCF      PORTA,1    ;turn off buzzer for 1/4 sec.
            CALL     DELAY
            BSF      PORTA,1    ;turn on buzzer for 1/4 sec.
            CALL     DELAY
            BCF      PORTA,1    ;turn off buzzer for 1/4 sec.
            CALL     DELAY
            BSF      PORTA,1    ;turn on buzzer for 1/4 sec.
            CALL     DELAY
            BCF      PORTA,1    ;Turn buzzer off.
            GOTO     BEGIN
END
```

Modifications to the dice project

Can you think of any modifications you can make to this program? Perhaps you could add a roll routine so that a few numbers are shown before the dice finally comes to rest on the number.

The initial display routine could also be customized.

You could throw a 7.

Dice using 12C508

The dice circuit used 8 outputs and 1 input a total of 9 I/O.

But LEDs 0 and 6, 1 and 5, 2 and 4 work in pairs, i.e. they are on and off together. If these LEDs were paralleled up, then we only need 6 I/O, e.g.:

- Input from Switch
- Output to Buzzer

- Output to LEDs 0 and 6
- Output to LEDs 1 and 5
- Output to LEDs 2 and 4
- Output to LED 3

This project can then be undertaken using the 6 I/O of the 12C508.

Project 2   Reaction timer

There are many question and answer games on the market that would benefit from a reaction timer which indicates the first player of a team to press. This project has the facility for up to 6 players.

The circuit diagram for this project illustrated in Figure 18.3 uses 6 inputs and 7 outputs.

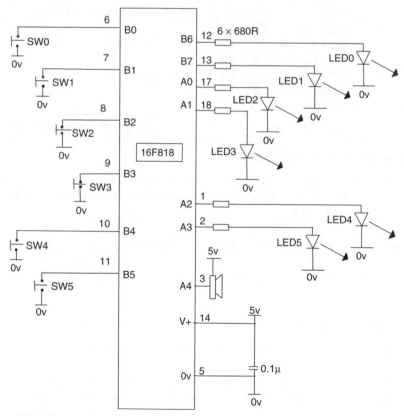

Figure 18.3 The reaction timer circuit

Reaction timer operation

If B0 is the first to press B6 output LED lights
If B1 is the first to press B7 output LED lights
If B2 is the first to press A0 output LED lights
If B3 is the first to press A1 output LED lights
If B4 is the first to press A2 output LED lights
If B5 is the first to press A3 output LED lights
The Buzzer is connected to A4.

The buzzer sounds for 4 seconds after a button is pressed. During this time no further presses are acknowledged. After the 4 seconds the buzzer stops and the LED is extinguished and the program resets.

The unit uses 13 I/O but not all 6 button/LED combinations need be used. The program will not need altering.

Just one point in case you were wondering: B0–B5 have been used as inputs instead of PORTA because PORTB has internal pull-up resistors on the inputs. The switches do not need their own – no point in using 5 resistors if you don't have to.

The reaction timer program

;REACTION.ASM

```
TMR0       EQU   1       ;means TMR0 is file 1.
STATUS     EQU   3       ;means STATUS is file 3.
PORTA      EQU   5       ;means PORTA is file 5.
PORTB      EQU   6       ;means PORTB is file 6.
ZEROBIT    EQU   2       ;means ZEROBIT is bit 2.
ADCON0     EQU   1FH     ;A/D Configuration reg.0
ADCON1     EQU   9FH     ;A/D Configuration reg.1
ADRES      EQU   1EH     ;A/D Result register.
CARRY      EQU   0       ;CARRY IS BIT 0.
TRISA      EQU   85H     ;PORTA Configuration Register
TRISB      EQU   86H     ;PORTB Configuration Register
OPTION_R   EQU   81H     ;Option Register
OSCCON     EQU   8FH     ;Oscillator control register.
COUNT      EQU   20H     ;COUNT a register to count events.
```

;***
;

```
          LIST        P=16F818    ;we are using the 16F818.
          ORG         0           ;the start address in memory is 0
          GOTO        START       ;goto start!
```

;**
;Configuration Bits

```
__CONFIG H'3F10'         ;sets INTRC-A6 is port I/O, WDT off, PUT on,
                         ;MCLR tied to VDD A5 is I/O
                         ;BOD off, LVP disabled, EE protect disabled,
                         ;Flash Program Write disabled,
                         ;Background Debugger Mode disabled, CCP
                         ;function on B2,
                         ;Code Protection disabled.
```

;**
;SUBROUTINE SECTION.

```
;0.1 second delay, actually 0.099968s
DELAYP1   CLRF        TMR0        ;START TMR0.
LOOPB     MOVF        TMR0,W      ;READ TMR0 INTO W.
          SUBLW       .3          ;TIME-3
          BTFSS       STATUS,
                      ZEROBIT     ;Check TIME-W = 0
          GOTO        LOOPB       ;Time is not = 3.
          NOP                     ;add extra delay
          NOP
          RETLW  0                ;Time is 3, return.

;4 second delay.
DELAY4    MOVLW       .40
          MOVWF       COUNT
LOOPC     CALL        DELAYP1
          DECFSZ      COUNT
          GOTO        LOOPC
          RETLW       0

;1 second delay.
DELAY1    MOVLW       .10
          MOVWF       COUNT
LOOPA     CALL        DELAYP1
          DECFSZ      COUNT
          GOTO        LOOPA
          RETLW       0
```

```
ON0      BSF      PORTB,6    ;Turn on LED0
         BSF      PORTA,4    ;Turn on buzzer
         CALL     DELAY4     ;Wait 4 seconds
         BCF      PORTB,6    ;Turn off LED0
         BCF      PORTA,4    ;Turn off buzzer
         GOTO     SCAN

ON1      BSF      PORTB,7    ;Turn on LED1
         BSF      PORTA,4    ;Turn on buzzer
         CALL     DELAY4     ;Wait 4 seconds
         BCF      PORTB,7    ;Turn off LED1
         BCF      PORTA,4    ;Turn off buzzer
         GOTO     SCAN

ON2      BSF      PORTA,0    ;Turn on LED2
         BSF      PORTA,4    ;Turn on buzzer
         CALL     DELAY4     ;Wait 4 seconds
         BCF      PORTA,0    ;Turn off LED2
         BCF      PORTA,4    ;Turn off buzzer
         GOTO     SCAN

ON3      BSF      PORTA,1    ;Turn on LED3
         BSF      PORTA,4    ;Turn on buzzer
         CALL     DELAY4     ;Wait 4 seconds
         BCF      PORTA,1    ;Turn off LED3
         BCF      PORTA,4    ;Turn off buzzer
         GOTO     SCAN

ON4      BSF      PORTA,2    ;Turn on LED4
         BSF      PORTA,4    ;Turn on buzzer
         CALL     DELAY4     ;Wait 4 seconds
         BCF      PORTA,2    ;Turn off LED4
         BCF      PORTA,4    ;Turn off buzzer
         GOTO     SCAN

ON5      BSF      PORTA,3    ;Turn on LED5
         BSF      PORTA,4    ;Turn on buzzer
         CALL     DELAY4     ;Wait 4 seconds
         BCF      PORTA,3    ;Turn off LED5
         BCF      PORTA,4    ;Turn off buzzer
         GOTO     SCAN
```

```
;************************************************************
;
```

;CONFIGURATION SECTION.

```
START       BSF         STATUS,5    ;Turns to Bank1.

            MOVLW       B'0000000'  ;8 bits of PORTA are O/P
            MOVWF       TRISA

            MOVLW       B'00000110' ;PORTA IS DIGITAL
            MOVWF       ADCON1

            MOVLW       B'00111111'
            MOVWF       TRISB       ;PORTB is mixed I/O

            MOVLW       B'00000000'
            MOVWF       OSCCON      ;oscillator 31.25kHz

            MOVLW       B'00000111' ;Prescaler is /256
            MOVWF       OPTION_R    ;TIMER is 1/32 secs.

            BCF         STATUS,5    ;Return to Bank0.
            CLRF        PORTA       ;Clears PortA.
            CLRF        PORTB       ;Clears PortB.
            CLRF        COUNT
```

;**

;Program starts now.
```
            MOVLW       0FFH
            MOVWF       PORTA       ;Turn on PORTA outputs
            BSF         PORTA,4     ;Turn on buzzer
            MOVWF       PORTB       ;Turn on PORTB outputs
            CALL        DELAY1      ;Wait 1 second
            CLRF        PORTA       ;Turn off PORTA outputs
            BCF         PORTA,4     ;Turn off buzzer
            CLRF        PORTB       ;Turn off PORTB outputs

SCAN        BTFSS       PORTB,0     ;Has B0 been pressed
            GOTO        ON0         ;Yes
            BTFSS       PORTB,1     ;Has B1 been pressed
```

```
         GOTO      ON1       ;Yes
         BTFSS     PORTB,2   ;Has B2 been pressed
         GOTO      ON2       ;Yes
         BTFSS     PORTB,3   ;Has B3 been pressed
         GOTO      ON3       ;Yes
         BTFSS     PORTB,4   ;Has B4 been pressed
         GOTO      ON4       ;Yes
         BTFSS     PORTB,5   ;Has B5 been pressed
         GOTO      ON5       ;Yes
         GOTO      SCAN
END
```

How does it work?

The program starts by turning all the LEDs and the buzzer on for 1 second to check they are all working.

The program then tests each input in turn starting with B0, if it is set i.e. not pressed the program skips and checks the next input. When the last input B5 is checked and it is not pressed then the program skips the next instruction and goes back to SCAN again.

If one of the inputs is pressed the program branches to the relevant subroutine to turn on the appropriate LED and buzzer for 4 seconds before returning to scan the switches again.

Reaction timer development

One way of making this program more interesting and to develop your programming skills – when a button is pressed have the outputs jump around B6, A0, A3, A1, A2 then B7 before landing on the correct output.

You could also have a flashing light routine at the start of the program to check they are working, you could also pulse the buzzer. The buzzer could be made to beep a number of times to give an audible indication of who was first to press. Another modification you could make is – think of one yourself, I'm not doing all the work.

Project 3   Burglar alarm

Operation

The circuit for the Burglar Alarm is shown in Figure 18.4 using the 16F818.

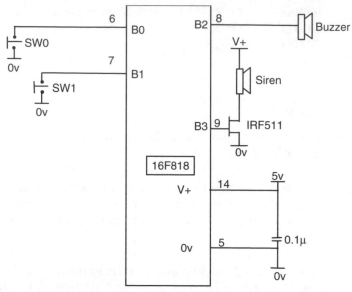

Figure 18.4 Burglar alarm circuit

It uses two inputs, SW0 and SW1 which are both normally closed. They can represent Door contacts, Passive Infra red sensor outputs, window contacts or tilt switches.

SW0 has a delay on it but SW1 is immediately active.

Both switches can have additional switches wired in series with them to provide extra security cover. If SW1 is a window contact in a caravan it could have a tilt switch wired in series with it, so if the caravan was moved the siren would sound immediately.

SW0 and SW1 are connected to PORTB so pull-ups are not required.

A buzzer is used to indicate entry and exit delays on the alarm and a siren is connected to the micro via an IRF511 (Power MOSFET).

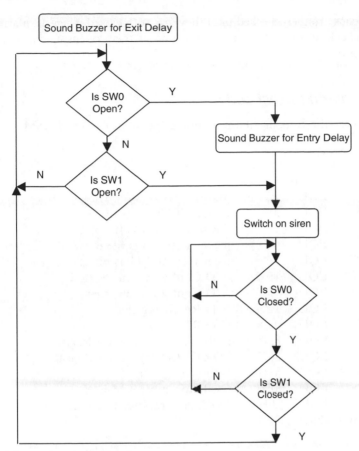

Figure 18.5 Burglar alarm flowchart

How does it work?

Consider the flow chart in Figure 18.5.

With reference to the flow chart:

When the alarm is switched on a 30 second exit delay is activated and the buzzer sounds for this time.

Switches 0 and 1 are continually checked until one of them is open.

If SW0 is opened a 30 second entry delay is activated and the buzzer sounds for this time, the siren will then sound for 5 minutes.

If SW1 is opened the siren will sound immediately for 5 minutes.

The switches are then checked until they are both closed when the alarm resets back to checking switches 0 and 1 until one of them opens again.

Switching off the power would disable the alarm.

Burglar alarm project code

The code for the Burglar Alarm is shown below in ALARM.ASM

;ALARM.ASM

```
;EQUATES SECTION
TMR0        EQU    1        ;means TMR0 is file 1.
STATUS      EQU    3        ;means STATUS is file 3.
PORTA       EQU    5        ;means PORTA is file 5.
PORTB       EQU    6        ;means PORTB is file 6.
ZEROBIT     EQU    2        ;means ZEROBIT is bit 2.
ADCON0      EQU    1FH      ;A/D Configuration reg.0
ADCON1      EQU    9FH      ;A/D Configuration reg.1
ADRES       EQU    1EH      ;A/D Result register.
CARRY       EQU    0        ;CARRY IS BIT 0.
TRISA       EQU    85H      ;PORTA Configuration Register
TRISB       EQU    86H      ;PORTB Configuration Register
OPTION_R    EQU    81H      ;Option Register
OSCCON      EQU    8FH      ;Oscillator control register.
COUNT       EQU    20H      ;COUNT a register to count events.
COUNTA      EQU    21H
```

```
;*********************************************************
        LIST      P=16F818   ;we are using the 16F818.
        ORG       0          ;the start address in memory is 0
        GOTO      START      ;goto start!
;*********************************************************
;Configuration Bits

__CONFIG H'3F10'           ;sets INTRC-A6 is port I/O, WDT off, PUT on,
                           ;MCLR tied to VDD A5 is I/O
                           ;BOD off, LVP disabled, EE protect disabled,
                           ;Flash Program Write disabled,
                           ;Background Debugger Mode disabled,
                           ;CCP function on B2,
                           ;Code Protection disabled.

;*********************************************************
```

;SUBROUTINE SECTION.

```
;0.1 second delay, actually 0.099968s
DELAYP1  CLRF    TMR0                    ;START TMR0.
LOOPB    MOVF    TMR0,W                  ;READ TMR0 INTO W.
         SUBLW   .3                      ;TIME-3
         BTFSS   STATUS,ZEROBIT          ;Check TIME-W = 0
         GOTO    LOOPB                   ;Time is not = 3.
         NOP                             ;add extra delay
         NOP
         RETLW 0                         ;Time is 3, return.


;0.5 second delay.
DELAYP5  MOVLW   .5
         MOVWF   COUNT
LOOPC    CALL    DELAYP1
         DECFSZ  COUNT
         GOTO    LOOPC
         RETLW   0


;1 second delay.
DELAY1   MOVLW   .10
         MOVWF   COUNT
LOOPA    CALL    DELAYP1
         DECFSZ  COUNT
         GOTO    LOOPA
         RETLW   0


;0.25 second delay
DELAYP25 MOVLW   .3
         MOVWF   COUNT
LOOPD    CALL    DELAYP1
         DECFSZ  COUNT
         GOTO    LOOPD
         RETLW   0


;5 second delay
DELAY5   MOVLW   .50
         MOVWF   COUNT
LOOPE    CALL    DELAYP1
         DECFSZ  COUNT
         GOTO    LOOPE
         RETLW   0
```

```
BUZZER    MOVLW     .5
          MOVWF     COUNTA    ;5 × 2 SECONDS
BUZZ1     BSF       PORTB,2
          CALL      DELAY1
          BCF       PORTB,2
          CALL      DELAY1
          DECFSZ    COUNTA
          GOTO      BUZZ1
          MOVLW     .10
          MOVWF     COUNTA    ;10 × 1 SECOND
BUZZ2     BSF       PORTB,2
          CALL      DELAYP5
          BCF       PORTB,2
          CALL      DELAYP5
          DECFSZ    COUNTA
          GOTO      BUZZ2
          MOVLW     .20
          MOVWF     COUNTA
BUZZ3     BSF       PORTB,2   ;20 × 0.5 SECONDS
          CALL      DELAYP25
          BCF       PORTB,2
          CALL      DELAYP25
          DECFSZ    COUNTA
          GOTO      BUZZ3
          RETLW     0
```

;***
;CONFIGURATION SECTION.

```
START     BSF       STATUS,5  ;Turns to Bank1.

          MOVLW     B'11111111'  ;8 bits of PORTA are I/P
          MOVWF     TRISA

          MOVLW     B'00000110'  ;PORTA IS DIGITAL
          MOVWF     ADCON1

          MOVLW     B'00000011'
          MOVWF     TRISB        ;PORTB is MIXED I/O

          MOVLW     B'00000000'
          MOVWF     OSCCON       ;oscillator 31.25kHz

          MOVLW     B'00000111'  ;Prescaler is /256
          MOVWF     OPTION_R     ;TIMER is 1/32 secs.
```

```
          BCF      STATUS,5   ;Return to Bank0.
          CLRF     PORTA      ;Clears PortA.
          CLRF     PORTB      ;Clears PortB.
          CLRF     COUNT
```

;**
;Program starts now.

```
          CALL     BUZZER     ;Exit delay
CHK_ON    BTFSC    PORTB,0    ;Check for alarm
          GOTO     ENTRY
          BTFSC    PORTB,1
          GOTO     SIREN
          GOTO     CHK_ON

ENTRY     CALL     BUZZER     ;Entry delay
SIREN     BSF      PORTB,3    ;5 minute siren
          MOVLW    .60
          MOVWF    COUNTA
WAIT5     CALL     DELAY5
          DECFSZ   COUNTA
          GOTO     WAIT5

          BCF      PORTB,3    ;Turn off Siren
CHK_OFF   BTFSC    PORTB,0    ;Check switches closed
          GOTO     CHK_OFF
          BTFSC    PORTB,1
          GOTO     CHK_OFF

          CALL     DELAYP25   ;antibounce
          GOTO     CHK_ON
END
```

The Burglar Alarm uses 2 inputs and 2 outputs a total of 4 I/O.

We can therefore program the Alarm with a 12C508 chip.

Burglar alarm using the 12C508

The circuit diagram for the Alarm with the 12C508 is shown in Figure 18.6.

Note in the circuit of Figure 18.6, showing the alarm using the 12C508, that no external oscillator circuit is required and that pull ups are not required on pins GPIO,0 or GPIO,1 (or GPIO,3). N.B. GPIO,3 is an input only pin.

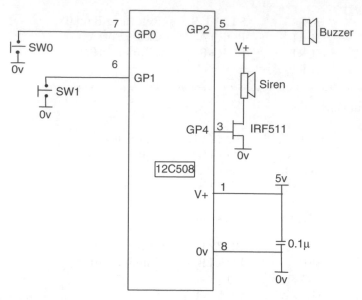

Figure 18.6 Burglar alarm using 12C508

The flowchart of course is the same. The code is shown below as ALARM_12.ASM using the header for the 12C508 from Chapter 15.

WARNING: The 12C508 only has a two level deep stack which means when you do a CALL you can only do one more CALL from that subroutine otherwise the program will get lost.

Program code for 12C508 burglar alarm

```
;ALARM_12.ASM FOR 12C508

TMR0        EQU        1           ;TMR0 is FILE 1.
GPIO        EQU        6           ;GPIO is FILE 6.
OSCCAL      EQU        5           ;Oscillator calibration.
STATUS      EQU        3           ;STATUS is FILE 3.
ZEROBIT     EQU        2           ;ZEROBIT is Bit 2.
COUNT       EQU        07H         ;USER RAM LOCATION.
TIME        EQU        08H         ;TIME IS 39
COUNTB      EQU        09H


;********************************************************************
```

```
          LIST        P=12C508    ;We are using the 12C508.
          ORG         0           ;0 is the start address.
          GOTO        START       ;goto start!
```

;**
;Configuration Bits

```
__CONFIG H'0FEA'        ;selects Internal RC oscillator, WDT off,
                        ;Code Protection disabled.
```

;**
;SUBROUTINE SECTION.

;1 second delay
```
DELAY1    MOVLW       .100                     ;100 × 1/100 SEC.
          MOVWF       COUNT
TIMEA     CLRF        TMR0                     ;Start TMR0
LOOPB     MOVF        TMR0,W                   ;Read TMR0 into W
          SUBWF       TIME,W                   ;TIME-W
          BTFSS       STATUS,ZEROBIT ;Check TIME-W=0
          GOTO        LOOPB
          DECFSZ      COUNT
          GOTO        TIMEA
          RETLW       0
```

;1/2 second delay
```
DELAYP5   MOVLW       .50                      ;50 × 1/100 SEC.
          MOVWF       COUNT
TIMEB     CLRF        TMR0                     ;Start TMR0
LOOPC     MOVF        TMR0,W                   ;Read TMR0 into W
          SUBWF       TIME,W                   ;TIME-W
          BTFSS       STATUS,ZEROBIT ;CHECK TIME-W=0
          GOTO        LOOPC
          DECFSZ      COUNT
          GOTO        TIMEB
          RETLW       0
```

;1/4 second delay
```
DELAYP25  MOVLW       .25                      ;25 × 1/100 SEC.
          MOVWF       COUNT
TIMEC     CLRF        TMR0                     ;Start TMR0
LOOPD     MOVF        TMR0,W                   ;Read TMR0 IN W
```

```
           SUBWF    TIME,W              ;TIME-W
           BTFSS    STATUS,ZEROBIT      ;Check TIME-W=0
           GOTO     LOOPD
           DECFSZ   COUNT
           GOTO     TIMEC
           RETLW    0
```

```
;2 second delay
DELAY2     MOVLW    .200                ;200 × 1/100 SEC.
           MOVWF    COUNT
TIMED      CLRF     TMR0                ;Start TMR0
LOOPE      MOVF     TMR0,W              ;Read TMR0 IN W
           SUBWF    TIME,W              ;TIME-W
           BTFSS    STATUS,ZEROBIT      ;Check TIME-W=0
           GOTO     LOOPE
           DECFSZ   COUNT
           GOTO     TIMED
           RETLW    0
```

```
BUZZER     MOVLW    .5
           MOVWF    COUNTB              ;5 × 2 Seconds
BUZZ1      BSF      GPIO,2
           CALL     DELAY1
           BCF      GPIO,2
           CALL     DELAY1
           DECFSZ   COUNTB
           GOTO     BUZZ1
           MOVLW    .10
           MOVWF    COUNTB              ;10 × 1 Second
BUZZ2      BSF      GPIO,2
           CALL     DELAYP5
           BCF      GPIO,2
           CALL     DELAYP5
           DECFSZ   COUNTB
           GOTO     BUZZ2
           MOVLW    .20
           MOVWF    COUNTB
BUZZ3      BSF      GPIO,2              ;20 × 0.5 Seconds
           CALL     DELAYP25
           BCF      GPIO,2
           CALL     DELAYP25
```

```
              DECFSZ    COUNTB
              GOTO      BUZZ3
              RETLW     0
```

;**
;CONFIGURATION SECTION.

```
START         MOVWF     OSCCAL
              MOVLW     B'00101011'  ;GPIO bits 2 and 4 are O/Ps.
              TRIS      GPIO
              MOVLW     B'00000111'
              OPTION                 ;PRESCALER is /256
              CLRF      GPIO         ;Clears GPIO
              MOVLW     .39
              MOVWF     TIME
```

;**

;Program starts now.

```
              CALL      BUZZER       ;Exit delay
CHK_ON        BTFSC     GPIO,0       ;Check for alarm
              GOTO      ENTRY
              BTFSC     GPIO,1
              GOTO      SIREN
              GOTO      CHK_ON


ENTRY         CALL      BUZZER       ;Entry delay
SIREN         BSF       GPIO,4       ;5 minute siren
              MOVLW     .150
              MOVWF     COUNTB
WAIT5         CALL      DELAY2       ;150 × 2 seconds
              DECFSZ    COUNTB
              GOTO      WAIT5
              BCF       GPIO,4       ;Turn siren off


CHK_OFF       BTFSC     GPIO,0       ;Check switches closed
              GOTO      CHK_OFF
              BTFSC     GPIO,1
              GOTO      CHK_OFF
              CALL      DELAYP25     ;antibounce
              GOTO      CHK_ON
END
```

Fault finding

What if it all goes wrong!

The block diagram of the microcontroller in Figure 18.7 shows 3 sections:

Inputs, the microcontroller and outputs.

Figure 18.7 Block diagram of the microcontroller circuit

The microcontroller makes the output respond to changes in the inputs under program control.

All microcontroller circuits will have outputs and most will have inputs.

Check the supply voltage

Check that the correct voltages are going to the pins. 5v on Vdd, pin 14 and MCLR, pin 4 and 0v on Vss, pin 5, on the 16F84.

Checking inputs

If the inputs are not providing the correct signals to the micro then the outputs will not respond correctly.

Before checking inputs or outputs it is best to remove the microcontroller from the circuit – with the power switched off. You have inserted the micro in an IC holder so that it can be removed easily! This is essential for development work.

In order to check the inputs and outputs to the microcontroller let us consider a circuit we have looked at before in Chapter 5, the Switch Scanning Circuit, shown below in Figure 18.8.

The four switches sw0, sw1, sw2 and sw3 turned on LED0, LED1, LED2 and LED3 respectively.

To test the inputs monitor the voltage on the input pins to the microcontroller, pins 1, 2, 17 and 18. They should go high and low as you throw the switches.

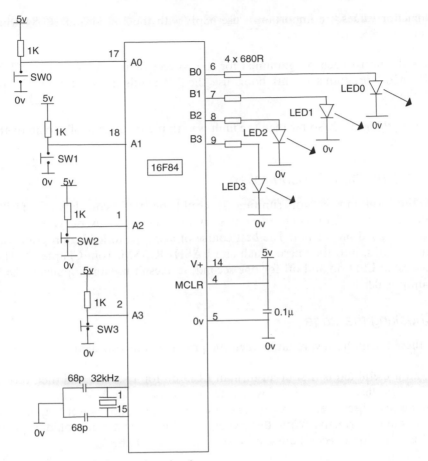

Figure 18.8 The switch scanning circuit

Checking outputs

The microcontroller will output 5v to turn on the outputs.

To make sure the outputs are connected correctly, apply 5v to each output pin in turn to make sure the corresponding LED lights.

When 5v is applied to pin 6, the B0 output then LED0 should light, etc. If it doesn't the resistor value could be incorrect or the LED faulty or in the wrong way round.

Check the oscillator

Check the oscillator is operating by monitoring the signal on CLKOUT, pin 15, with an oscilloscope or counter. Correct selection of the oscillator capacitor values are important – use 68pF with the 16C54 and 16F84 when using a 32kHz crystal.

Has the micro been programmed for the correct oscillator: R-C, LP, XT or HS. Most programs in this book use the LP configuration for the 32kHz Oscillator.

If everything is OK so far then the fault is with the microcontroller chip or the program.

Checking the microcontroller

If the program is not running it could be that you have a faulty microcontroller. You could of course try another, but how do you know if that is a good one or not. The best course of action is to load a program you know works, into the micro. Such as FLASHER.ASM from Chapter 2. This flashes an LED on and off for one second, it doesn't use any inputs and only 1 output B0.

Checking the code

If there are no hardware faults then the problem is in your code.

I find a useful aid is first of all turn an LED on for 1 second and then turn it off. When this works you know that the microcontroller is ok, and that your timing has been set correctly and the oscillator and power supply are functioning correctly. With the switch scanning circuit you could turn all 4 LEDs on for 1 second anyway to serve as an LED check.

To check your code, break it up into sections. Look at were the program stops running to identify the problem area.

If possible turn on LEDs on the outputs to indicate where you are in the program. If you are supposed to turn LED3 on when you go into a certain section of code and LED3 doesn't turn on, then of course you have not gone into that section you are stuck somewhere else.

These instructions can be removed later when the program is working.

Using a simulator

By using a simulator such as the one contained in MPLAB you can single step through the program and check it out a line at a time. To use the simulator from MPLAB select – Debugger, Select Tool, MPLAB SIM as shown in Figure 18.9.

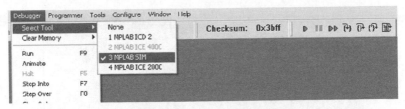

Figure 18.9 Selecting MPLAB SIM

Common faults

Here are just a few daft things my students (or I!) have done:

- Not switched the power on.
- Put the chip in upside down.
- Programmed the wrong program into the micro.
- Corrected faults in the code but forgot to assemble it again, thus blowing the previous incorrect HEX file again.
- Programmed incorrect fuses, i.e. Watchdog Timer and Oscillator.

Development kits

There are a number of development kits on the market (and you can make your own). They have a socket for your micro, inputs and outputs that you can connect to your micro. They are ideal for program development. Once verified using the kit if the system does not work then your circuit is at fault. I have developed such a kit shown in Figure 18.10. Details of it can be found on the SL Electrotech website at:
http://www.slelectrotech.com

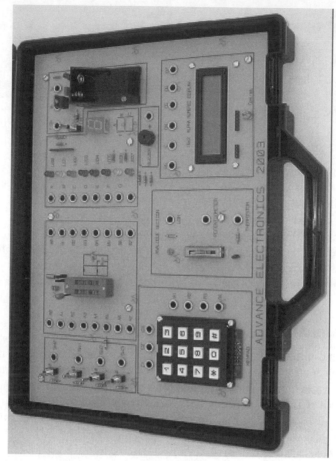

Figure 18.10 PIC microcontroller development kit

19
Instruction set, files and registers

Microcontrollers work essentially by manipulating data in memory locations. Some of these memory locations are special registers others are user files. In a control application data may be read from an input port, manipulated and passed to an output port.

To use the microcontroller you need to know how to move and manipulate this data in the memory. There are 35 instructions in the PIC 16F84 to enable you to do this. Using the Microcontroller is then about using these instructions in a program. Like any vocabulary you do not use all the words all of the time, some you never use others only now and again. The PIC Instruction Set is like this – you can probably manage quite well with say 15 instructions.

Most of these instructions involve the use of the WORKING REGISTER or Wreg. The W register is at the heart of the PIC Microcontroller. To move data from File A to File B you have to move it from File A to W and then from W to File B, rather like a telephone system routes one caller to another via the exchange. The W reg also does the arithmetic and logical manipulating on the data.

The PIC microcontroller instruction set

To communicate with the PIC microcontroller you have to learn how to program it using its instruction set. The 16F84 chip has a 1k × 14 bit word EEPROM program memory, 68 × 8bit general purpose registers and a 35 word instruction set made up of three groups of instructions, bit, byte and literal and control operations.

The instructions can be sub-divided into 3 types:

- Bit Instructions, which act on 1 bit in a file.
- Byte Instructions, which act on all 8 bits in a file.
- Literal and Control Operations, which modify files with variables or control the movement of data from one file to another.

Bit instructions

The bit instructions act on a particular bit in a file, so the instruction would be followed by the data which specifies the file number and bit number.

I.e. BSF 6,3 This code is not too informative so we would use something like BSF PORTB,BUZZER where PORTB is file 6 and the buzzer is connected to bit 3 of the output port. In the equates section we would see PORTB EQU 6 and BUZZER EQU 3.

BCF	Bit Clear in File.
BSF	Bit Set in File.
BTFSC	Bit Test in File Skip if Clear.
BTFSS	Bit Test in File Skip if Set.

Byte instructions

Byte instructions work on all 8 bits in the file. So a byte instruction would be followed by the appropriate file number.

I.e. DECF 0CH. This statement is not too informative so we would again indicate the name of the file such as DECF COUNT. Of course we would need to declare in the equates section that COUNT was file 0CH, by COUNT EQU 0CH.

ADDWF	ADD W and F.
ANDWF	AND W and F.
CLRF	CLeaR File.
CLRW	CLeaR Working register.
COMF	COMplement File.
DECF	DECrement File.
DECFSZ	DECrement File Skip if Zero.
INCF	INCrement File.
INCFSZ	INCrement File Skip if Zero.
IORWF	Inclusive-OR W and F.
MOVF	MOVe F to W.
MOVWF	MOVe W to F.
NOP	No OPeration.
RLF	Rotate File one place Left.
RRF	Rotate File one place Right.
SUBWF	SUBtract W from F.
SWAPF	SWAp halves of F.
XORWF	eXclusive-OR W and F.

Literal and control operations

Literal and control operations manipulate data and perform program branching (jumps).

ADDLW   ADD Literal with W.
ANDLW   AND Literal with W.
CALL   CALL subroutine.
CLRWDT   CLeaR watchdog Timer.
GOTO   GOTO address.
IORLW   Inclusive-OR Literal with W.
MOVLW   MOVe Literal to W.
RETFIE   RETurn From IntErrupt.
RETLW   RETurn place Literal in W.
RETURN   RETURN from subroutine.
SLEEP   Go into standby mode.
SUBLW   SUBtract Literal from W.
XORLW   eXclusive-OR Literal and W.

These instructions operate mainly on two 8 bit registers – the Working register or W register and a File F which can be one of the 15 special registers or one of the 68 general purpose file registers which form the user memory (RAM) of the 16F84.

The memory map of the 16F84 is shown in Figure 6.1.

The PIC Microcontrollers are 8 bit devices – this means that the maximum number that can be stored in any one memory location is 255. Some PICs like the 17C43 have 454 bytes of data memory. So to address memory locations greater than 255 the idea of pages or Banks has been introduced. Bank0 holds address locations up to 255, while Bank1 can hold a further 255 and Bank2 a further 255 etc. So you need to know what Bank a particular register or file is in.

Banks are not used in the 16C54.

Registers

Registers are made up of 8 bits as shown in Figure 19.1.

bit 7	bit 6	bit 5	bit 4	bit 3	bit 2	bit 1	bit0
1	0	1	1	0	0	1	0

MSB --- LSB

Figure 19.1 Register layout

Bit 0 is the Least Significant Bit (LSB) and Bit 7 is the Most Significant Bit (MSB).

Register 00 indirect data addressing register

See File Select Register, Register 04.

Register 01 TMR0, TIMER 0/counter register

This register can be written to or read like any other register. It is used for counting or timing events. The contents of the register can be incremented (add 1) by the application of an external pulse applied to the TOCKI pin i.e. counting cars into a car park or by the internal instruction cycle clock which runs at ¼ of the crystal frequency to time events.

Register 02 PCL, program counter

The Program Counter automatically increments to execute program instructions. An application of the use of the Program Counter is illustrated in the section on the Look Up Table, in Chapter 8.

Register 03, status register

The Status Register contains the result of the arithmetic or logical operations of the program. The 8 bits of the Status Register are shown in Figure 19.2.

bit 7	bit 6	bit 5	bit 4	bit 3	bit 2	bit 1	bit0
IRP	RP1	RP0	TO	PD	Z	DC	C

Figure 19.2 Status Register

- Bit 0, C, Carry Bit. This is (set to a 1) if there is a carry from an addition or subtraction instruction.
  E.g. if one 8 bit number is added to another;

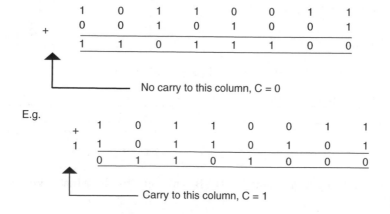

If the result of a subtraction is +ve or zero then the carry bit is set.

If the result of a subtraction is −ve then the carry bit is clear.

- Bit 2, Z, Zero Bit. This is set if the result of an arithmetic or logic operation is zero. i.e. countdown to zero.
  An important use of this bit is checking if a variable in memory is equal to a fixed value. I.e. does file CARS contain 150.

```
MOVLW   .150            ;Put 150 in W
SUBWF   CARS,W          ;Subtract W from CARS, i.e. CARS-150
BTFSS   STATUS,ZEROBIT  ;Zerobit set if CARS = 150
```

- Bits 6 and 5, RP1 and RP0, are the bank select bits to address banks 0,1,2 and 3 to select the different registers and user files.
  00 would select bank0, 01 selects bank1, 10 selects bank2 and 11 selects bank3.

Register 04 FSR file select register

The file select register is used in conjunction with the Indirect Data Addressing Register, Register 00. They are used in indirect addressing to read or write data not from a specific file, but to or from a file indicated by the data in the file select register.

Register 05 PORT A and register 06 PORT B

Ports are the pin connections that allow the microcontroller to communicate with its surroundings. Port A is a 5 bit port on the 16F84, only the 5 LSB's are used. Port A bit0 can also be programmed to be a clock input (T0CKI). Port B is an 8 bit port. To set up a port the instruction TRIS is used. Tris is an abbreviation for tristate, three states which can be a high impedance input, a high (5v) output or a low (0v) output.

Register 8FH oscillator control register (16F818)

The oscillator control register is used to select the clock frequency when using the internal oscillator.

bit7	bit6	bit5	bit4	bit3	bit2	bit1	bit0
-	IRCF2	IRCF1	IRCF0	-	IOFS	-	-

Figure 19.3 Oscillator control register

bit 6–4   IRCF2:IRFC0: Internal Oscillator Frequency Select Bits.
$\qquad$ $111 = 8$ MHz (8MHz source drives clock directly)
$\qquad$ $110 = 4$ MHz
$\qquad$ $101 = 2$ MHz
$\qquad$ $100 = 1$ MHz
$\qquad$ $011 = 500$ kHz
$\qquad$ $010 = 250$ kHz
$\qquad$ $001 = 125$ kHz
$\qquad$ $000 = 31.25$ kHz (INTRC source drives clock directly)

bit2   IOFS:INTOSC Frequency Stable Bit.

W Register

The W register holds the result of an operation or an internal data transfer. It is like a telephone exchange – data comes into the W register and is transferred out to another file.

Option Register

This register is used to prescale the Real Time Clock/Counter. TMR0 clock runs at ¼ of the crystal frequency but can be divided down by the prescaler for longer time measurements.

Stack

Stack is the name given to the memory location that keeps track of the program address when a Call instruction is made. There is an eight level stack in the 16F84, which means that the program can jump to a subroutine and from there jump to another subroutine, making 8 jumps in total and the stack will be able to return it back to the program. The 16C54 has a two level stack.

Instruction set summary

ADDLW   Adds a number (literal) to W.
$\qquad$ E.g. ADDLW 7 will add 7 to W, the result is placed in W.

ADDWF   Adds the contents of W to F.
E.g. ADDWF 7 will add the contents of the W register and file 7
N.B. the result is placed in file 7.
E.g. ADDWF 7,W the result is placed in W.
Status affected C,DC,and Z.

ANDLW   The contents of W are ANDed with an 8 bit number (literal).
The result is placed in W.
E.g. ANDLW 12H or ANDLW B'00010010' or ANDLW .18
Status affected Z.

ANDWF   The contents of W are ANDed with F.
E.g. ANDWF 12,W the contents of file 12 is ANDed to the
contents of W. N.B. The result is placed in W.
E.g. ANDWF 12 the result is placed in file 12.
Status affected Z.

BCF   Clear the bit in file F.
E.g. BCF 6,4 bit 4 is cleared in file 6.
File 6 is port B this clears bit 4, i.e. bit $4 = 0$.

BSF   Set bit in file F.
E.g. BSF 6,4 this sets bit 4 in File 6, i.e. bit $4 = 1$.

BTFSC   Test bit in file skip if clear.
E.g. BTFSC 3,2 this tests bit 2 in file 3 if it is clear then the
next instruction is missed. File 3 is the status register bit 2 is the
zero bit so the program jumps if the result of an instruction was
zero.

BTFSS   Test bit in file skip if set.
E.g. BTFSS 3,2 if bit 2 in file 3 is set then the next instruction is
skipped.

CALL   This calls a subroutine in a program.
E.g. CALL WAIT1MIN This will call a routine (you have
written) to wait for 1 minute. May be to turn a lamp on for 1
minute, and then return back to the program.

CLRF   This clears file F i.e. all 8 bits in file F are cleared.
E.g. CLRF 5.
Status affected Z.

CLRW   This clears the W register.
            Status affected Z.

CLRWT   The watchdog timer is cleared. The watchdog is a safety device in
            the microcontroller if the program crashes the watchdog timer
            times out then restarts the program.
            Status affected TO, PD.

COMF   The 8 bits in file F are complemented i.e. inverted.
            E.g. COMF 6.
            Status affected Z.

DECF   Subtract 1 from file F. Useful for counting down to zero.
            E.g. DECF 12 will store the result in 12.
            DECF 12,W will store the result in W leaving 12 unchanged.
            Status affected Z.

DECFSZ   The contents of F are decremented and the next instruction is
            skipped if the result is zero.
            E.g. DECFSZ 12 or DECFSZ COUNT

GOTO   This is an unconditional jump to a specified location in the
            program.
            E.g. GOTO SIREN.

INCF   Add 1 to F. This value could then be compared to another to see
            if a total had been achieved.
            E.g. INCF 14 or INCF COUNT
            Status affected Z.

INCFSZ   Add 1 to F if the result is zero then skip the next instruction.
            E.g. INCFSZ 19 or INCFSZ COUNT

IORLW   The contents of the W register are ORed with a literal.
            E.g. IORLW 27.

i.e W	=	1	0	0	1	1	0	1	1
L	=	0	0	0	1	1	0	0	1
L + W	=	1	0	0	1	1	0	1	1

This is a very useful way of determining if any bit in a file
is set i.e. by ORing a file with 00000000 if all the bits in the

file are zero the OR result is zero and the zero bit is set in the status register.
Status affected Z.

IORWF   The contents of the W register are ORed with the file F.
E.g. IORWF 7,W The result is stored in W.
E.g. IORWF 7 The result is stored in file 7.
Status affected Z.

MOVF   The contents of the file F are moved into the W register, from there the data can be moved to an output port.
E.g. MOVF 12,W File 12 is moved to W.
E.g. MOVF 12 File 12 is moved to file 12? Zero is affected.
Status affected Z.

MOVLW   The 8 bit literal is moved directly into W.
E.g. MOVLW .127
Status affected Z.

MOVWF   The contents of the W register are moved to F.
E.g. MOVWF 6 the data in the W register is placed on port B.

NOP   No operation – may seem like a daft idea but it is very useful for small delays. The NOP instruction delays for ¼ of the clock speed.

OPTION   The contents of W are loaded into the OPTION register. This instruction is used to prescale i.e. set TMR0 timing rate as shown in Figure 19.4.

RETFIE   This instruction is used to return from an interrupt.

RETLW   This instruction is used at the end of a subroutine to return to the program following a CALL instruction. The literal value is placed in the W register. This instruction can also be used with a look up table.
E.g. RETLW 0

RETURN   This instruction is used to return from a subroutine.

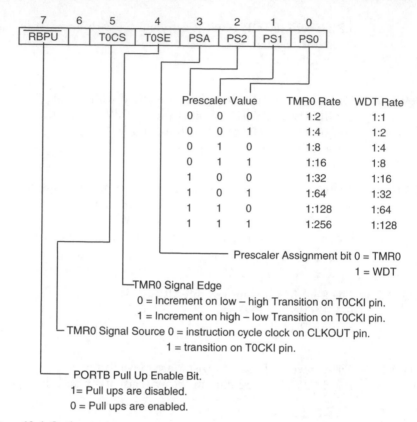

Figure 19.4 Option register

RLF   The contents of the file F are rotated 1 place to the left through the carry flag. Shifting a binary number to the left means that the number has been multiplied by 2. This instruction is used when multiplying binary numbers.

E.g. RLF 12,W The result is placed in W.

E.g. RLF 12 The result is placed in file 12.

The diagram below shows file 12 being rotated left.

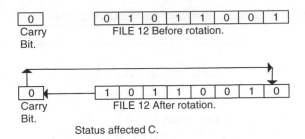

RRF   This is the same as RLF except the file is rotated one place to the right.

SLEEP   When executing this instruction the chip is put into sleep mode. The power-down status bit (PD) is cleared, the time-out status bit is set, the watchdog timer and its prescaler are cleared and the oscillator driver is turned off. The watchdog timer still keeps running from its own internal clock.
E.g. SLEEP
Status affected TO, PD.

SUBLW   The contents of the W register are subtracted from a number.
E.g. SUBLW 14 executes 14-W the result is placed in W. The carry bit and the zero bit in the status register are affected
N.B.   If W > 14 then C = 0 the result is   ve.
             If W < 14 then C = 1 the result is +ve or zero.
             If W = 14 then Z = 1 the result is zero. This is
a very useful condition. To find out if something has
occurred 14 times subtract 14 from those occurrences
if the answer is zero – bingo.
Status affected C, DC, and Z.

SUBWF   The contents of the W register are subtracted from the contents of the file F.
E.g. SUBWF 14,W executes F-W the result is placed in W.
E.g. SUBWF 14 executes F-W the result is placed in F.
NB.   If W > F then C = 0 the result is −ve.
             If W < F then C = 1 the result is +ve or zero.
             If W = F then Z = 1 the result is zero.
Status affected C, DC, and Z.

SWAPF   The upper and lower nibbles (4 bits) of file F are swapped.
E.g. SWAPF 12,W The result is placed in W.
E.g. SWAPF 12 The result is placed in file 12.

File 12 before SWAPF

0	1	1	0	1	1	0	1

File 12 after SWAPF

1	1	0	1	0	1	1	0

TRIS   Load the TRIS register.
            The contents of the W register are loaded into the TRIS register.
            This then configures an I/O port as input or output.
            E.g. MOVLW   B'00001111'
                 MOVWF   TRISB
            This sets the 4 LSB's of port B as inputs and the 4 MSB's as
            outputs. N.B. 1 for an input, 0 for an output.

XORLW   The contents of the W register are Exclusive Ored with the
            literal. If the result is zero then the contents match.
            i.e. If a number on the input port, indicating temperature, is the
            same as the literal then the result is zero and the zero bit is set.
            i.e. $0 \oplus 0 = 0$, $0 \oplus 1 = 1$, $1 \oplus 0 = 1$, $1 \oplus 1 = 0$.
            E.g. XORLW 67
            Status affected Z.

XORWF   The contents of the W register are Exclusive Ored with the
            contents of the file F. i.e. If a number on the input port,
            indicating temperature, is the same as the W register then the
            result is zero and the zero bit is set. N.B. you can not Exclusive
            OR the input port directly with a file, you have to do this by
            loading the file into the W register with an MOVF instruction.
            E.g. XORWF 17,W The result is placed in W.
            E.g. XORWF 17 The result is placed in 17.
            Status affected Z.

Did you notice how vital the W register is in the operation of the
microcontroller?

Data cannot go directly from A to B, it goes from A to W and then from
W to B.

Appendix A
Microcontroller data

Product	Program Memory Bytes (words)	EEPROM Data Memory Bytes	RAM Bytes	I/O Pins	A/D Channels	Timers	Max Speed MHz	Internal Oscillator MHz
12C508	768 (512)	-	25	6	-	1–8 bit	4	4
12C509	1536 (1024)	-	41	6	-	1–8 bit	4	4
12CE518	768 (512)	16	25	6	-	1–8 bit	4	4
12CE519	1536 (1024)	16	41	6	-	1–8 bit	4	4
12CE673	1792 (1024)	16	128	6	4 (8 bit)	1–8 bit	4	10
12CE674	3584 (2048)	16	128	6	4 (8 bit)	1–8 bit	4	10
12F629	1792 (1024)	128	64	6	-	1–8 bit 1–16 bit	20	4
12F675	1792 (1024)	128	64	6	4 (10 bit)	1–8 bit 1–16 bit	20	4

Product	Program Memory		E²Prom Data Memory	RAM Bytes	8-Bit ADC Channels	I/O Ports	Timers	MAX Speed MHz
	Bytes	Words						
PIC16CXXX – 4-12 Interrupts, 200ns Instruction Execution, 35 Instructions, 4MHz Internal Oscillator, 4/5 Oscillator Selections								
PIC1F83	896	512 × 14	64	36	-	13	1-8bit, 1-WDT	10
PIC16F84	1792	1024 × 14	64	68	-	13	1-8bit, 1-WDT	10
PIC16F872	3584	2048 × 14	64	128	5 (10 bit)	22	1-16bit, 2-8bit, 1-WDT	20
PIC16F873	7168	4096 × 14	128	192	5 (10 bit)	22	1-16bit, 2-8bit, 1-WDT	20
PIC16F874	7168	4096 × 14	128	192	8 (10 bit)	33	1-16bit, 2-8bit, 1-WDT	20
PIC16F876	14336	8192 × 14	256	368	5 (10 bit)	22	1-16bit, 2-8bit, 1-WDT	20
PIC16F877	14336	8192 × 14	256	368	8 (10 bit)	33	1-16bit, 2-8bit, 1-WDT	20
PIC16C923	7168	4096 × 14	-	176		52	1-16bit, 2-8bit, 1-WDT	8
PIC16C924	7168	4096 × 14	-	176	5	52	1-16bit, 2-8bit, 1-WDT	8
PIC17CXXX – 4-12 Interrupts, 200ns Instruction Execution, 35 Instructions, 4MHz Internal Oscillator, 4/5 Oscillator Selections								
PIC17C42A	4096	4096 × 14	-	192	8	33	1-16bit, 2-8bit, 1-WDT	20
PIC17C43	8192	8192 × 14	-	368	5	22	1-16bit, 2-8bit, 1-WDT	20
PIC17C44	16384	8192 × 14	-	368	8	33	1-16bit, 2-8bit, 1-WDT	20
PIC17C752	16384	2048 × 14	-	256	6 (12 bit)	16	1-16bit, 2-8bit, 1-WDT	20
PIC17C756	32768	4096 × 14	-	256	6 (12 bit)	16	1-16bit, 2-8bit, 1-WDT	20
PIC17C762	16384	4096 × 14	-	256	6 (12 bit)	22	1-16bit, 2-8bit, 1-WDT	20
PIC17C766	32768	4096 × 14	-	256	10 (12 bit)	33	1-16bit, 2-8bit, 1-WDT	20
PIC18CXXX – 10 MIPS, 77 Instructions, C-compiler Efficient Instruction Set, Table Operation, Switchable Oscillator Sources								
PIC18C242	16384	8192 × 16	-	512	5 (10 bit)	23	3-16bit, 2-8bit, 1-WDT	40
PIC18C442	16384	8192 × 16	-	512	8 (10 bit)	34	3-16bit, 2-8bit, 1-WDT	40
PIC18C252	32768	1634 × 16	-	1536	5 (10 bit)	23	3-16bit, 2-8bit, 1-WDT	40
PIC18C452	32768	1634 × 16	-	1536	8 (10 bit)	34	3-16bit, 2-8bit, 1-WDT	40

Appendix B
Electrical characteristics

Absolute maximum ratings: (16F818/9)

Absolute maximum ratings: (16F818/9)	
Ambient temperature	$-55°C$ to $+125°C$
Storage temperature	$-65°C$ to $+150°C$
Voltage on any pin with respect to Vss (except Vdd and MCLR)	$-0.6V$ to Vdd $+0.6V$
Voltage on Vdd with respect to Vss	0 to $+7.5V$
Voltage on MCLR with respect to Vss	0 to $+14V$
Total power dissipation	1W
Max. current out of Vss pin	200mA
Max. current into Vdd pin (16C54)	50mA
Max. current into Vdd pin	200mA
Max. output current sunk by any I/O pin	25mA
Max. output current sourced by any I/O pin	25mA
Max. output current sourced by PORTA	100mA
Max. output current sourced by PORTB	100mA
Max. output current sunk by PORTA	100mA
Max. output current sunk by PORTB	100mA

DC Characteristics.

PIC12F629/675

Characteristic	Symbol	Min.	Typ.	Max.	Units	Conditions.
Supply Voltage	Vdd					Fosc = DC to 4MHz
		2.0		5.5	V	With A/D off
		2.2		5.5	V	PIC12F675 withA/D on
		3.0		5.5	V	Fosc = 4 to 10MHz
RAM dataretention voltage	Vdr	1.5			V	Device in Sleep Mode
Supply Current	Idd		0.4	2	mA	Fosc = 4MHz, Vdd = 2V
			0.9	4	mA	Fosc = 4MHz, Vdd = 5.5V
			5.2	15	mA	Fosc = 20MHz, Vdd = 5.5V
			20	48	µA	Fosc = 32KHz, Vdd = 2V, WDT disabled.
Power down Current (sleep mode)	Ipd		1	18	µA	Vdd = 2.0V, A/Don
			0.9		µA	Vdd = 2.0V, WDT disabled

PIC16F818/9

Characteristic	Symbol	Min.	Typ.	Max.	Units	Conditions.
SupplyVoltage	Vdd	2.0		5.5	V	HS, XT, RC and LP osc modes
RAM dataretention voltage	Vdr	1.5			V	Device in Sleep Mode
Supply Current	Idd		28		µA	Fosc = 32KHz, Vdd = 5.0V
			874		µA	Fosc = 4MHz, Vdd = 5.0V
Power down Current (sleep)	Ipd		0.5		µA	Vdd = 5.0V

PIC16F84

Characteristic	Symbol	Min.	Typ.	Max.	Units	Conditions.
Supply Voltage	Vdd					
PIC16F84-XT		4.0		6.0	V	
PIC16F84-RC		4.0		6.0	V	
PIC16F84-HS		4.5		5.5	V	
PIC16F84-LP		4.0		6.0	V	
RAM dataretention voltage	Vdr	1.5			V	Device in Sleep Mode
Supply Current	Idd					
PIC16F84-XT			7.3	10	mA	Fosc = 4MHz, Vdd = 5.5V
PIC16F84-RC			7.3	10	mA	Fosc = 4MHz, Vdd = 5.5V
PIC16F84-HS			5	10	mA	Fosc = 10MHz, Vdd = 5.5V
PIC16F84-LP			35	400	µA	Fosc = 32KHz, Vdd = 3.0V, WDT disabled.
Power down Current (sleep mode)	Ipd		40	100	µA	Vdd = 4.0V, WDT enabled
			38	100	µA	Vdd = 4.0V, WDT disabled

PIC16F87X

Characteristic	Symbol	Min.	Typ.	Max.	Units	Conditions.
Supply Voltage	Vdd	4.0		5.5	V	LP, XT, RC osc configuration
		4.5		5.5	V	HS osc configuration
RAM dataretention voltage	Vdr	1.5			V	Device in Sleep Mode
Supply Current	Idd		1.6	4	mA	Fosc = 4MHz, Vdd = 5.5V
			7	15	mA	Fosc = 20MHz, Vdd = 5.5V
			20	35	µA	Fosc = 32KHz, Vdd = 3.0V, WDT disabled.
Power down Current (sleep)	Ipd		1.5	19	µA	Vdd = 4.0V, WDT enabled

Appendix C
Decimal, binary and hexadecimal numbers

Homosapiens are used to Decimal numbers, i.e. 0,1,2,3......9. Electronic machines or chips use Binary numbers 0 and 1, (OFF and ON).

Decimal numbers increase in tens, i.e. 267 means 7 ones, 6 tens and 2 hundreds.

$$100 \quad 10 \quad 1$$
$$2 \quad \ \ 6 \quad \ 7$$

Binary numbers increase in twos, i.e. 1010. The right hand 0 means no ones, the next digit means 1 two, the next means no fours, the next 1 eight etc.

$$8 \quad 4 \quad 2 \quad 1$$
$$1 \quad 0 \quad 1 \quad 0$$

The binary number 1010 consists of 4 *BI*nary digi*T*s it is called a 4 BIT number. 1010 is equivalent to 10 in decimal numbers.

We can change decimal numbers to binary and binary numbers to decimal. Digital systems, i.e. Computers are a little better than we are at this.

Consider the decimal number 89, to turn this into a binary number write the binary scale:

$$128 \quad 64 \quad 32 \quad 16 \quad 8 \quad 4 \quad 2 \quad 1$$

To make 89 we need $(0 \times 128) + (1 \times 64) + (0 \times 32) + (1 \times 16) + (1 \times 8) + (0 \times 4) + (1 \times 2) + (1 \times 1)$.

So 89 in decimal = 01011001 in binary.

To convert a binary number to decimal add up the various multiples of 2, i.e. 10011010 is:

$$128 \quad 64 \quad 32 \quad 16 \quad 8 \quad 4 \quad 2 \quad 1$$
$$1 \quad \ \ 0 \quad \ \ 0 \quad \ \ 1 \quad \ 1 \quad 0 \quad 1 \quad 0$$
$$= 128 + 16 + 8 + 2 = 154.$$

A long string of binary numbers is difficult to read, i.e. 11010101 to make this shorter and therefore easier to put into a microcontroller Hexadecimal numbers are used. Hexadecimal numbers increase in sixteen's and are described by sixteen digits. Table C.1 shows these 16 digits and their decimal and binary equivalents.

Table C.1 4 BIT Decimal, binary and hexadecimal representation

Decimal	Binary	Hexadecimal
0	0000	0
1	0001	1
2	0010	2
3	0011	3
4	0100	4
5	0101	5
6	0110	6
7	0111	7
8	1000	8
9	1001	9
10	1010	A
11	1011	B
12	1100	C
13	1101	D
14	1110	E
15	1111	F

The PIC microcontrollers are 8 bit micros, they use 8 binary digits for number representation like
10010101 this is

$$128 \quad 64 \quad 32 \quad 16 \quad 8 \quad 4 \quad 2 \quad 1$$
$$1 \quad\ \ 0 \quad\ \ 0 \quad\ \ 1 \quad\ 0 \quad 1 \quad 0 \quad 1$$
$$= 149$$

The largest decimal number that can be represented by an 8 bit number is: 11111111 which represents:-

$$128 \quad 64 \quad 32 \quad 16 \quad 8 \quad 4 \quad 2 \quad 1$$
$$1 \quad\ \ 1 \quad\ \ 1 \quad\ \ 1 \quad\ 1 \quad 1 \quad 1 \quad 1$$
$$= 255$$

But we can program our microcontroller to increase our number representation from 8 bits i.e. up to 255:
to 16 bits, numbers up to 65,535
to 24 bits, numbers up to 16,777,215
to 32 bits, numbers up to 4,294,967,295 etc.

As mentioned earlier hexadecimal numbers are a shorter way of writing binary numbers. To do this divide the binary number into groups of 4 and write each group of 4 as a hex number.

i.e. 10010110 as 1001 0110 in binary

$$= \quad 9 \quad 6 \quad \text{in} \quad \text{hex.}$$

i.e. 11011010 as 1101 1010 in binary

$$= \quad D \quad A \quad \text{in} \quad \text{hex.}$$

Table C.2 shows some of the 255 numbers represented by 8 bits.

Table C.2 8 BIT Decimal, binary and hexadecimal representation

Decimal	Binary	Hexadecimal
0	00000000	00
1	00000001	01
2	00000010	02
3	0000011	03
4	00000100	04
5	00000101	05
8	00001000	08
15	00001111	0F
16	00010000	10
31	00011111	1F
32	00100000	20
50	00110010	32
63	00111111	3F
64	01000000	40
100	01100100	64
127	01111111	7F
128	10000000	80
150	10010110	96
200	11001000	C8
250	11111010	FA
251	11111011	FB
252	11111100	FC
253	11111101	FD
254	11111110	FE
255	11111111	FF

Appendix D
Useful contacts

- Author
d.w.smith@mmu.ac.uk

- A Microcontroller Design Company
S.L. Electrotech Limited.
☎+44(0) 782 566626 http://www.slelectrotech.com

- Arizona Microchip, the company that manufacture the PICs. This Website is a must.
http://www.MICROCHIP.COM

- Places to buy your components
Farnell ☎+44(0) 113 263 6311 http://www.Farnell.com
Rapid Electronics ☎+44(0) 1206 751166
RS Components ☎+44(0) 1536 444105 http://www.rs-components.com/rs
Maplin Electronics ☎+44(0) 1702 554000 http://www.maplin.co.uk

- A recommended Magazine
Everyday Practical Electronics
http://www.epemag.wimborne.co.uk

Index